COAL

Energy and chemical storehouse

by

Dr. I. BERKOVITCH
F.R.I.C., M.I.Chem.E.

1978

PORTCULLIS PRESS

Previous book by the same author
"Coal on the Switchback" published by George Allen & Unwin,
1977.

Computerset and printed in Gt Britain by Page Bros (Norwich)
Ltd, Mile Cross Lane, Norwich.

COAL
Energy and chemical storehouse

Preface

At what level of technical understanding do you pitch a 'popular science' book? I take it that the people likely to read this one have an interest in technical subjects including some basis in chemistry. This basis may be but a dim memory of the meanings of the terms 'elements' and 'molecules' with perhaps further vague recollections of the simpler compounds and of chemical reactions. But that will be enough. It relieves me of the need, in a relatively short book, of discussing these basics or the nature of chemical changes.

So my starting point is to assume this earlier technical interest by the reader. On this basis we have a look at the energy scene, consider chemicals and their current sources before moving on to the probable pattern of the future—the 're-discovery' of coal as a major storehouse of both energy and chemicals with reserves for some centuries ahead. For this renewed role for coal is now being increasingly recognised by the industrial world as it faces up to realising that the other fossil fuels have reserves only for decades, that nuclear power has many problems and that alternative sources will take decades to establish on any material scale.

Acknowledgments

I am very glad to acknowledge the support I have had in preparing this book from the National Coal Board and in particular from Geoffrey Kirk, Sheila Conchie, Peter Paul, Mary Barnes, Ray Jennings, Peter Heap, John Scott and Keith Beeston. Naturally, they are not responsible for the use I made of their highly-valued advice and assistance.

Contents

Preface   v

Chapter 1   Re-discovering coal   1
World Trends — trends in the UK — re-discovering coal

Chapter 2   Chemicals and their sources   14
Inorganics — organics — some familiar derivatives — vary-
ing the route — prospects for supplies — technical appendix

Chapter 3   Chemical character of coal   33
Composition — classifying coal — as rank increases

Chapter 4   Chemicals from coal — past and present   46
When coal is carbonized — coal processing — coal into
gas — the largest chemicals from coal plant — metals from
coal? — current developments

Chapter 5   Energy from coal — past and present   59
Cleaning up the atmosphere — using waste heat — can fuel
cells save us fuel? — what of MHD? — how efficiently do we
use fuel?

Chapter 6   Fluidization — versatile technique   78
Are there any drawbacks? — burning coal — working at
higher pressures — other applications

Chapter 7   Chemicals and fuels from coal — the developing   93
future
Pyrolysis — extracting with liquid solvents — extracting with
super-critical gas — hydrogenation — gasification — underground gasification — building up chemicals — does it pay to
convert coal? — basic science is not neglected

Chapter 8   Coalplexes — making the maximum use of coal   115
Possible forms of Coalplex — the integrated project COAL-
COM — COALCON and its difficulties — linking with nu-
clear heating — prospects

Chapter 9   The Coal International   125
ECSC — ECE Coal Committee — IEA Coal Projects —
Other agreements

Chapter 10   Prospects in the UK   132
Government support — What is happening in the great wide
world? — UK Future

Index   139

Foreword

Not more than five years ago, the average citizen could expect to go through life with no more knowledge of energy resources, or of coal conversion processes, than that required to answer a question in some school examination. The potential of nuclear power seemed unlimited; the flow of cheap oil was unrestricted and assumed to be inexhaustible. The coal industry might be contracting, but that was no bad thing as coal was a bit 'old-fashioned' and the concept of the 'coal tree' with aspirin, sulphonamide drugs and nylons suspended from its branches was rather overshadowed by the much greater importance of that lusty new giant industry, petrochemicals. All of these impressions have been assailed by subsequent events and our average citizen has had to grapple with the insecurity of oil supplies, the problems besetting the nuclear power industry and the possible contributions coal might make to alleviate these problems, not to mention the economic consequences—inflation and balance of payments crises. He has need, therefore, of guidance in the form of a readable account of the background to the energy situation and the important role of coal in helping to ease the problems of the future. I believe this book can meet that requirement.

J. Gibson
Member for Science
National Coal Board

Chapter 1

Re-discovering Coal

Coal is the most abundant fossil fuel in the world, and in Britain. Once the main source of energy and of chemicals in all the industrialised countries, coal was increasingly superseded in both these roles by oil and gas after the Second World War. But the shock of the enormous price rises of oil in 1973 did more than simply restore former King Coal to being a fuel once again competitive with oil. By placing new huge burdens on the import costs of the industrialised countries, it transformed energy supply from being a narrow esoteric subject studied only by its own specialists to one of great public interest. For example, it is difficult to imagine a Secretary of State for Energy feeling justified in any previous period in calling a public conference of interested parties on energy policy as Tony Benn did in June 1976.

And the combination of price rise and embargo on oil supplies—even if relatively short-lived—directed attention in all the major oil-using countries to both security of energy supply and to assessing reserves of the various possible fuels. Further important off-shoots were growing interest in raising efficiency of using fuel and power, in renewable sources of energy—solar, wind, tides and waves—in virtually inexhaustible ones such as geothermal and others which might extend our reserves such as breeder nuclear reactors.

For Britain, the perspective—clearly set out in a discussion document prepared in mid-1976 for the Advisory Council on Research and Development for Fuel and Power (ACORD)—was seen in these terms: even at Britain's previous average annual rate of growth in primary energy consumption of less than 2 per cent, consumption would double by 2010 A.D. North Sea oil and gas should bring important economic benefits but production seemed likely to peak before the end of the 1980's. Not long afterwards a gap would begin to develop between UK total energy demand and its indigenous supplies.

1

From the analysis of various possible 'scenarios' for the future, the first implication in respect of primary energy was that considerable attention needed to be paid to coal utilisation and conversion technologies. Before we discuss this important document, let us first turn to the background of energy developments on a world scale and see the features and changes that have led to the present 're-discovery' of coal.

World Trends

Over the thirty years or so before 1972, the world had been 'consuming' energy at a rate that increased on the average by a little over five per cent annually. In the decade before the sharp oil price rises, the rate was actually a little faster—about $5\frac{1}{2}$ per cent. This corresponded to an increase of energy usage per head at the rate of about $3\frac{1}{2}$ per cent a year. Joel Darmstadter and Sam H. Schurr, reporting these findings in 1973, went on to point to the marked shifts in regional shares and variations in regional growth rates. The Soviet Union's share of world energy consumption (only 2 per cent in 1925) grew from 10 per cent just before the Second World War to 15 per cent in 1970. Japan in that same post-war period shot up from 2·3 per cent to 5·2, while the USA though retaining its leading world share fell back from 45 per cent to less than a third of the world total. The study also revealed the gross differences between energy use per head in the industrialised world and the poor countries; the most extreme disparity was between North America on the one hand and the levels in most of Africa and the developing countries of Asia; the ratio was forty to one. Between the USA and other industrialised areas such as Western Europe, the ratio was about $2\frac{1}{2}$ (in absolute figures 329 million Btu per head per year against 134).

There had also been the most marked shift in proportion of different sources of energy. In the two decades following 1950 oil and gas between them rose from accounting for 38 per cent of world energy to 64 per cent; coal over the same period declined from 56 per cent to 29 per cent. The move away from emphasis on coal occurred in all the principal regions of the world. In Western Europe there was a sharp decline in the relative share of coal but also in the absolute quantity. Even the two coal giants of Western Europe—West Germany and the UK—showed heavy decreases of coal output. However USA, Eastern Europe and the USSR slightly raised their use of coal at average rates of between one and two per cent per annum, while China also increased at a rather faster rate. The overall effects of these various trends is summarised in the table below, slightly abbreviated from one by Darmstadter and Schurr.

Thus, the integrated effect on a world scale was of a small continuing

Table 1. World Energy Consumption. Average annual rates of increase by source, 1950 to 1970.

Period	1950–1970 %	1960–1970 %
Coal	2·3	2·0
Oil	7·6	8·0
Natural gas	9·0	8·6
Primary electricity	5·2	5·7
TOTAL	5·3	5·6

From paper by Darmstadter and Schurr at meeting 'Energy in the 1980s' The Royal Society, London, November 1973.

increase in using coal accompanied by very much larger rates of growth for both oil and natural gas resulting in the growing shift in proportion noted earlier. However in considering these figures the relative contribution made by each of these sources must naturally be taken into account. From the same survey, the distribution of energy sources is given as follows:

Table 2. World Energy Consumption. Distribution of Energy Sources.

Year	1950 %	1970 %
Coal	55·7	31·2
Oil	28·9	44·5
Natural gas	8·9	17·8
Primary electricity	6·5	6·5

Over this period the absolute figure for energy consumed was estimated to have risen from 77 quadrillion (US) Btu to 214 quadrillion (US) Btu.

These changes are evident from the various diagrammatic presentations. Figure 1 is a graph of world primary energy consumption, while Fig. 2 gives a block diagram showing the changes for the last two decades. The marked differences between different parts of the world are indicated in Fig. 3 and an alternative presentation of the changes in source in Fig. 4.

A further important feature was the transformation of the USA from a nett oil exporter to a heavy importer of oil. In 1962 imports accounted for 20 per cent of the use of oil. Despite the price jolt, the patriotic exhortations of successive Presidents, Project Independence and establishment of new government organisations to promote this independence,

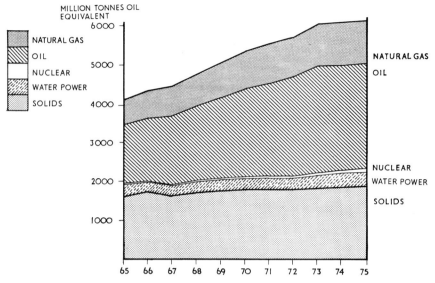

Figure 1. World Primary Energy Consumption. From BP statistical review of the world oil industry 1975. (taking 1 tonne oil = 1·5 tonnes coal).

the proportion of imports continued to rise and was about half of total consumption in 1977. Domestic supplies of natural gas also began to level off, then decline. Consequently imports of liquefied natural gas also grew, further increasing the energy dependence of the USA on imports. Within the EEC coal fell from its primary position after the war to supplying about one fifth of the energy needs in 1975 (with a further

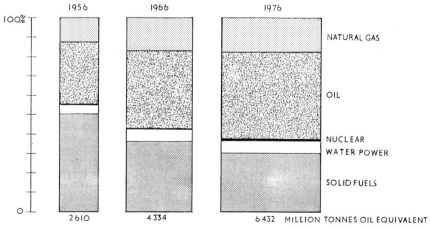

Figure 2. World Primary Energy Consumption 1956, 1966 and 1976. From BP statistical review of the world oil industry 1976.

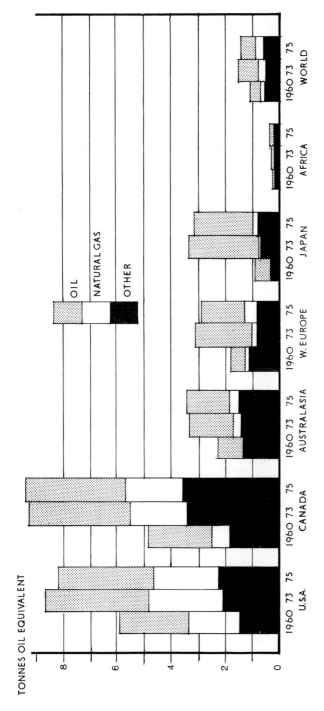

Figure 3. Energy Consumption per head 1960, 1973 and 1975. From BP statistical review of the world oil industry 1975 (taking 1 tonne oil = 1·5 tonnes coal).

contribution of 3 per cent from lignite); oil provided well over a half, natural gas a sixth and 6 per cent came from primary electricity. In the same year the Netherlands closed its last pit. The Commission pointed out that the EEC as a unit had become the world's biggest oil importer and in 1974 proposed a strategy of aiming to reduce dependence on imported energy from 60 per cent to 50 per cent by 1985.

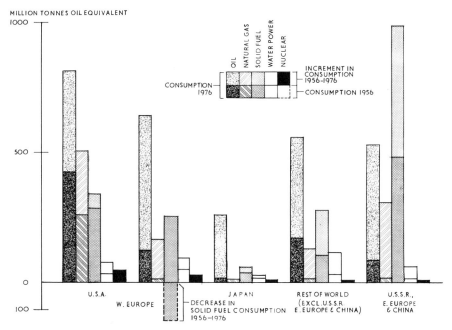

Figure 4. Changes in Primary Energy Consumption 1956–1976. From BP statistical review of the world oil industry 1976 (taking 1 tonne oil = 1·5 tonnes coal).

Early hopes of major supplies of nuclear energy proved to be exaggerated. By 1975 the USA had 57 commercial plants 'on line' providing only a little over nine per cent of electrical requirements, and the reality within the EEC fell short of expectations. 'Technical and environmental problems' led to a shortfall in the nuclear programme so that the share of total gross electricity produced was 6 per cent in 1974. In Britain, Mr. Arthur Hawkins, Chairman of the CEGB, has referred to our first nuclear programme as 'highly successful', with Britain at the end of 1974 accounting for almost a third of all world generation of nuclear power outside the Soviet Union. Yet the amount of nuclear generated electricity stayed at much the same level, with fluctuations, for some eight years and in 1976 represented about a tenth of the units generated. There have been great difficulties in bringing the second generation of reactors, the Advanced Gas Reactors (AGR), into commission but two AGR stations were operating in 1977 and a further three were expected to come on stream by 1979. Over the period to 1982 official estimates were that nuclear electricity would double its share of the market for primary energy use.

Trends in the UK

The waxing and waning in importance of the various primary fuels in the UK is clearly shown in the graph and charts (Figs 5 and 6). Over the

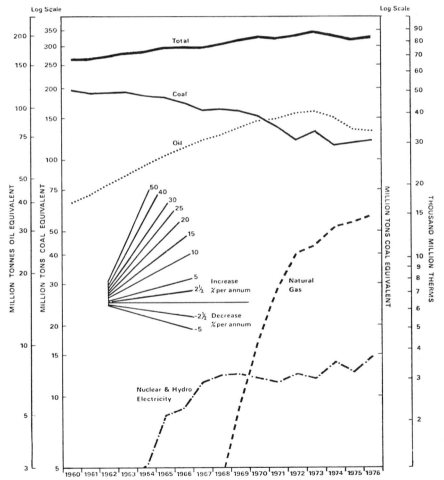

Figure 5. Total Inland Consumption of Primary Fuels. From 'Digest of UK Energy Statistics 1977' HMSO.

Percentage Shares

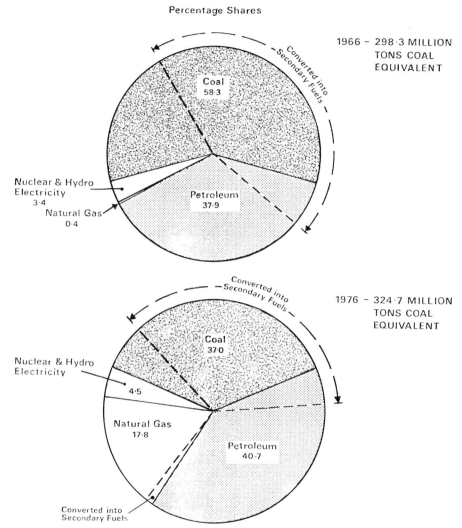

1966 – 298·3 MILLION
TONS COAL
EQUIVALENT

Coal
58·3

Converted into
Secondary Fuels

Nuclear & Hydro
Electricity
3·4

Natural Gas
0·4

Petroleum
37·9

1976 – 324·7 MILLION
TONS COAL
EQUIVALENT

Converted into
Secondary Fuels

Coal
37·0

Nuclear & Hydro
Electricity

4·5

Natural Gas
17·8

Petroleum
40·7

Converted into
Secondary Fuels

Figure 6. Total Inland Consumption of Primary Fuels 1966 and 1976. From 'Digest of UK Energy Statistics 1977' HMSO.

period from before the start of these diagrams and up to 1973 petroleum grew in importance, while coal steadily declined. Natural gas had a meteoric rise in its first few years from 1968 then settled down to a smaller rate of increase (though it should be noted that the vertical scale is on a logarithmic basis where the interval 5 to 50 is equal in spacing to that from 20 to 200 which foreshortens the higher values and exaggerates the smaller ones in presentation). The contribution from hydro-electricity has remained fairly steady and the rising amounts of nuclear-generated electricity up to 1968 are as clear as the lack of increase in the following years.

The peak of coal consumption for the post-war period was reached in 1956. Coal had earlier been cheaper than oil in pence per therm, off-setting the advantages of the liquid fuel in higher calorific value and easier handling but their prices as fuel for power stations became equal in 1958; thereafter, apart from the imposition of a fuel tax in 1961, oil remained cheaper until the startling days of the early 1970s.'During this period in the 1960s the West European Coal Producers organisation and notably Dr E. Schumacher, then Chief Economist of the NCB, were actively pointing to the dangers in the increasing dependence on oil. Demand for oil was growing faster than future ability to supply; consequently, they argued, its current cheapness was deceptive and must be transient since the bargaining position of the oil exporters would strengthen and prices rise.

Against this, the 1967 White Paper on Fuel Policy claimed that we had become a four fuel economy able to draw on natural gas and nuclear power as well as petroleum and coal. Consequently in respect of oil, it was

'right to base fuel policy on the expectation that regular supplies will continue to be available . . . difficult to predict the course of oil prices. There are a number of reasons for expecting them not to increase . . . it seems likely that oil will remain competitive with coal, and that pressure to force up crude oil prices will be held in check by the danger of loss of markets.'

For coal the policy led to the conclusion that further decline in markets could not be prevented unless the level of coal protection was raised progressively; this, it was argued, would be undesirable since

'excessive protection for coal would lead to a misallocation of man-power and capital to the detriment of the economy as a whole.'

So, with some degree of buffering against the effects of the social dislocation, the coal industry was to be run down, losing manpower at a rate of 35,000 a year.

However, in 1973, the bargain that was oil was drastically re-priced. What were the consequences in energy policy?

Re-discovering coal

Reactions in user-countries to both the sharply increased cost of oil, and to the feeling of uncertainty engendered by the embargo used as a political weapon, have had many facets; they include

* renewed emphasis on energy saving ranging from improving insulation to fresh interest in thermal linkage scheme;
* re-assessment of energy sources previously found uneconomic, such as tar sands and oil shales;
* re-costing of the products of conventional but poorer sources of fossil fuel that were not competitive at earlier price levels;
* new examination of geothermal and renewable energy resources—biomass (plants, micro-organisms), sewage, city refuse, agricultural waste, solar, wind, tidal and wave—now transmuted from 'crank' projects to mainstream, orthodox research;
* notably in USA and Western Europe, fresh attention to coal, once again made competitive by courtesy of OPEC;
* increased advocacy of nuclear reactors since they yield large amounts of energy from small quantities of fissile fuel, and notably breeder reactors;
* a new wave of research on nuclear fusion which could provide energy without radioactive waste.

It is possible that all of this can be comprised within the compass of one basic idea and its consequences—that of considering reserves of fossil fuels nation by nation and on a world scale, relating them to the rates of use and considering the consequences. This was certainly true of the big users of energy. But statements from several of the large oil exporters show that they too had become aware of the transience of their present stores of wealth. This is likely to have been a factor in the dizzy escalation of oil prices in 1973, alongside the stated one of using a weapon against those aiding their enemies and the implicit one of taking commercial advantage of having supplies indispensable to the industrialised world. Now the oil-supplying states are feverishly seeking to industrialise their countries to create a longer-lasting basis for future prosperity.

For the UK oil import costs leapt from £712 million in 1970 to almost £4,000 million in 1974. Mr Denis Healey, the Chancellor, observed in his Autumn 1974 Budget speech—reasonably enough—that we were going to be forced to alter our behaviour. The Department of Energy had earlier—under the previous administration—been separated from the Department of Trade and Industry; the official announcement explained that the re-organisation had been made since 'energy problems arising from developments in the supply and price of oil had become so particularly acute as to call for a special concentration, at both Ministerial and Official level, of time and effort.' NEDO published results of studies on the implications for British industry of the increased cost of energy, and the Central Policy Review Staff—popularly known as the 'think-tank'—issued its own report on energy conservation.

In a major policy speech in mid-1974, the then Secretary of State for Energy, Eric Varley, emphasised the need to develop our 'substantial' indigenous resources. This involved an examination of the coal industry, review of policy towards North Sea resources and consideration of nuclear reactor choice. Alongside these studies we needed also to take a 'much stronger interest now in the other side—energy use'. An Energy Technology Support Unit was established for special studies of new sources and also a number of consultative and advisory committees linking leading industrial figures with academics and civil servants. Several forms of international co-operation were developed, notably the International Energy Agency intended to secure oil supplies in emergency situations, and to build up co-operation among consumers; part of its stated objectives was also to promote co-operation among producers and consumers. The coal technology R & D projects set up under its auspices, where Britain plays a leading role, are discussed later in Chapter 9.

The combination of economic and policy pressures reduced consumption of oil faster than that of coal in the years 1974 to 1976 and occasionally—as in March 1975—resulted in coal edging out oil as the major fuel in the UK for the first time since 1970.

Looking towards the future, the discussion document on energy research and development (briefly mentioned on page 1) warned of the

'danger that the prospect of self-sufficiency, albeit for a limited period, could cause complacency . . . This would be a grave mistake . . . We must . . . take advantage of this breathing space given to us by North Sea oil to improve the efficiency of energy use and to develop, prove and establish those technologies upon which this country's future is likely to depend'.

To provide the necessary basis for the constructive public comment that they invited, ACORD examined possible views of the future that they termed 'scenarios'. They were briefly defined by the phrases

* trends continued;
* low growth;
* limit on nuclear;
* high energy cost;
* sharp rise in prices in 1990s;
* self-sufficiency;
* high growth.

The first of the implications of these scenarios in terms of primary energy was seen as 'considerable attention' to coal utilisation and conversion technologies. After referring to developing Continental Shelf technologies and those for deepwater finds of oil, to expanding nuclear technology with effective safety measures and R & D on alternative sources of energy, ACORD returned to coal and the need for new technologies for producing from it substitute natural gas and hydrocarbon liquids.

Naturally the policy implications included also attention to improving the efficiency of converting primary energy to electricity, and developing economic ways of saving energy.

From these implications, research priorities and the options for implementing policies were outlined. It is a truism that all are important to future economic growth. But within the area of the main subject matter of this book, what is significant is the rediscovery of coal in this country—as Aneurin Bevan put it—'built on coal' where nevertheless official policy so recently was to reduce the size of the industry.

The Final Report of the coal industry examination in 1974, approved by the government, 'pointed the way to a new future for coal'. Made up of representatives of the government, the NCB and unions of miners and managers, the 'tripartite group' endorsed the NCB's Plan for Coal recommending investment to yield new capacity for 42 million tons a year. But it also emphasised the importance of research in developing new uses for coal, and a Progress Report three years later reiterated the commitment to a substantial R & D effort. It is this area we shall be examining in following chapters after first considering the importance of chemicals, their current sources and what we are already getting from coal.

Further Reading

'Energy in the 1980s'. Philosophical Transactions of the Royal Society, Vol. 276, No. 1261, 1974.

Energy R & D in the UK. A discussion document prepared for the Advisory Council on Research and Development for Fuel and Power. Energy Paper No. 11, Department of Energy, HMSO 1976.

'Britain's nuclear workhorses'. Address to European Nuclear Conference 21 April 1975 by Mr. Arthur Hawkins, Chairman of CEGB, CEGB 1975.

Digest of UK Energy Statistics 1976, HMSO, 1976.

Fuel Policy. Cmnd 3438, HMSO, 1967.

Speech by Mr. Eric Varley, Secretary of State for Energy to the Association of British Science Writers, Department of Energy, June 26, 1974.

'The increased cost of energy—implications for UK industry'. National Economic Development Office, London, HMSO, 1974.

'Energy Conservation'. A report by the Central Policy Review Staff. London, HMSO, 1974.

Coal Industry Examination. Final Report 1974. Department of Energy.

'The Rediscovery of Coal' by Sir Derek Ezra, Chairman of the National Coal Board. The Spectator, December 13, 1975 (reprinted by the NCB).

Department of Energy. Report on research and development, 1974–75, HMSO, 1975

Department of Energy. Report on research and development, 1975–76, HMSO, 1976

'Coal on the Switchback. The Coal Industry since Nationalisation'. Israel Berkovitch, George Allen & Unwin, 1977.

'Coal for the Future—Progress with Plan for Coal and Prospects to the Year 2000'. Department of Energy, Millbank, London, 1977.

Fact Sheet No. 4. 'UK Coal Industry'. Department of Energy, Millbank, London, 1977.

Chapter 2

Chemicals and their sources

What is there in common between fertilisers, plastics, paints and pills, dry-cleaning solvents, soaps and synthetic rubber, or polishes, pigments and pesticides? These products all share the feature of being based on the output of the chemical and allied industries. In the industrialised countries, chemicals are major contributors to the economy, though virtually all the products are intermediates, supplied to other manufacturers for making final consumer goods. The buyer of the plastic bowl is not likely to have seen the plastic itself nor the mixed moulding powder from which the bowl is formed; nor will the do-it-yourself home painter be familiar with the resins, pigments, extenders, plasticisers and solvents that may have gone into his bright gloss paint.

Yet chemicals are among the fastest growing industries. On a world scale it was estimated that production grew at 10·1 per cent per annum over the period 1963–73, while the figure for all industries was 6·2; for the UK, the same comparison showed chemical industries as 6·6 and all industries as 3·2, so that the chemical industries were expanding at more than double the average rate. A more recent check on the figures showed an even more striking relationship. With output in 1970 as the base (100 per cent), chemical industries' output in the UK rose to 128·7 for 1976. The comparable statistic for all industrial production was 102·3. Early in 1975, Martin Trowbridge, director-general of the Chemical Industries Association (CIA) summarised its achievements in these terms:

'. . . the industry is the second largest in Europe. It deploys about 16 per cent of UK manufacturing industry's investment, employs 5 per cent of the UK workforce, and produces directly nearly 10 per cent of its net output.'

When its main customers are other industries, why do chemicals grow

14

so much faster? Mr Trowbridge summarises the issue as due to these trends:

'. . . chemical industry products have been progressively substituted for traditional raw materials (synthetic fibres replacing natural fibres; plastics replacing metals, timber, and paper/board products) either due to price, availability, superior properties or a combination of these. Also new uses have been found for chemical products either as manufactured goods for consumption or as 'enabling' supplies in the manufacturing processes.'

A CIA survey in 1976 revealed that their affiliated companies had adopted investment programmes that would give capacity for further increases averaging 7 per cent a year up to 1980. An important feature is that the chemical industries invest heavily in research and have a high rate of innovation. Both processes and products tend to change more frequently than in other industries. Improvements are constantly being offered to customers for resins, pesticides, pharmaceuticals, rubbers and the rest of the wide range of chemical products. Both the markets and the principal producers are highly international; plant has been designed generally in ever larger units demanding high capital investment, although there is now some discussion on whether limits have been reached in economies of scale in some cases.

Nevertheless, over a wide range, there is a clear reduction in unit cost as size of plant increases. Shell for example have published a table showing that the total cost of ethylene per tonne falls from £58 for a plant of 100,000 tonnes capacity a year to £37 for one of five times the capacity; this is on the basis of a constant cost of feedstock per unit, constant prices for products and full load operation.

The chemical producers understandably complain that some of their weaker brethren cut prices in order to maintain demand and keep the mammoth plants working at high loading. There is a complex interaction of raw material cost, product prices, growth of markets and technical progress. Its effect for the two decades ending in 1972 proved to be of direct benefit to consumers. For instance the published price of low density polyethylene in the UK—bucking the trend for almost all other products—fell over that period from a little over £300 per tonne to about half that figure. Meanwhile consumption soared from a negligible amount to over 350,000 tonnes a year.

Naturally in the following years, these trends were overtaken by the effects of the dramatic increase in the prices of oil—a major raw material for chemicals—then by those of widespread inflation and in due course the world recession. Consequently chemicals have finally followed the pattern of other products in becoming dearer, although at a lower rate.

In practice chemicals are likely to go through three principal stages of manufacture before they reach the consumer.

1. Those generally called 'base chemicals'—an ambiguous expression since the word 'base' also has a different meaning in pure chemistry—are made on a very large scale that may range up to millions of tons a year in the world as a whole. Among them—transcending the chemical division mentioned below—are the acids and alkalies, and also organic compounds, notably ethylene, propylene, butadiene, benzene, toluene and xylenes.

2. From these base chemicals are derived intermediates. And there are often many intermediate stages in converting materials before they are of service for the third stage. Among the intermediates are solvents, many industrial chemicals, polymers.

3. Finally the intermediates are chemically converted or physically blended or processed into the final products; these are the drugs, cosmetics, and other direct consumer products, or the myriad proprietary products, the mixed moulding powders, cleaning mixtures, additives and the rest supplied to industry.

A very important broad division of chemicals is into the categories of 'organic' and 'inorganic'. This distinction arose historically because of the false assumption that the compounds of carbon in living organisms could be produced only by a vital force. After this was disproved and carbon compounds were synthesised outside living things, the name 'organic' was nonetheless retained for the major branch of chemistry dealing with most of the carbon compounds. The inorganics are very largely based on minerals but again there are important exceptions. Let us consider the inorganic compounds first.

Inorganics

Output of the materials covered by the industry sector group defined as inorganics in the UK in 1974 was valued at £550 million. Though they are much less relevant to the main theme of this book, it is useful for completeness to mention here briefly some examples of the kinds of materials, their uses and the sources of their raw materials.

The inorganics serve other parts of the chemical industry, then iron and steel, textiles, building materials and other industries. They are considered 'mature products' offering little scope for substitution in the way indicated in the quotation from Trowbridge. Consequently the chemical Economic Development Committee * considered that their development

*(of the National Economic Development Council—NEDC—in 1976)

was likely to be at much the same rate as industry as a whole and their pattern not likely to alter. The largest product group is that of acids and alkalies.

Sulphuric acid—still sometimes taken as a measure of economic activity—is used in making superphosphate fertiliser, rayon, in cleaning metals before plating, producing explosives, dyestuffs, pigments and many other chemicals as well as going into electric batteries. Soda ash has so many uses that in giving them it is difficult to avoid the appearance of a catalogue; among them are making glass, ceramics, soap, refining petroleum, processing metals, treating pulp and paper. Cyanides (inorganic compounds of carbon), formerly so beloved of murder story writers for dispatching their victims, are extensively applied for the more innocuous duties of electroplating, heat treating metals, cleaning metals, fumigating and as intermediates in making many other chemicals.

Then there is oxygen, now increasingly used to increase the capacity of steel and iron furnaces, or with fuel gases for cutting and welding metals; medical grades are applied in anaesthesia and in helping patients in other ways. Dipping further at random into a mixed bag of materials we find for example magnesium oxide valuable in insulation for high temperatures, in pharmaceuticals and cosmetics, specialist cements, and in a food grade as an alkali and neutraliser.

Raw materials for these products—many of low unit value—are often available within the UK. They include salt, limestone, dolomite, air and sea water. Some minerals however have to be imported such as bauxite, borax and notably sulphur, where well over a million tons have to be brought in each year. Though sulphur can be obtained from pyrites and the sulphate of indigenous anhydrite, the use of the element sulphur—unfortunately meaning imports—is technically preferred. In the UK, the quantities recovered in purification processes are small and there appears to be little in our oil and gas, though elsewhere these are important sources.

Above I mentioned that there are important exceptions to the use of solids removed from the earth as sources of inorganics. A major one is the case of ammonia. Though there is a small contribution to supplies in the form of by-product ammonium sulphate from carbonising coal, most of our production, like most of that in the world, is based on reaction between steam and a hydrocarbon—generally using natural gas. The reaction yields carbon monoxide, carbon dioxide and hydrogen; further reaction of this carbon monoxide with additional steam yields more hydrogen and converts the former almost wholly to the dioxide which is then removed leaving the hydrogen needed for the ammonia synthesis. Air is earlier introduced into the reaction mixture. After the reaction its nitrogen remains and proportions are so controlled that the final gas contains the correct ratio of three volumes hydrogen to one of nitrogen needed to form the ammonia.

Another of the exceptions is recovering valuable salts from the seas—a notable case being the Dead Sea. This has vast reserves of magnesium chloride, common salt, potassium chloride and magnesium bromide which are being extracted in rising amounts. On a world scale, it was estimated recently that over 60 per cent of all magnesium and of all bromine were produced from sea-water.

Organics

In the other great division of chemicals—the organics—compounds have an essential structure of carbon atoms, very often combined with hydrogen but in many cases with other elements, with or without hydrogen.

Let us turn then to the organic base chemicals.

Though organic chemicals were earlier made only from vegetable and animal materials and from coal, and these remain important sources even today, over 90 per cent of the current world production of organics comes ultimately from petroleum oil and gas. Coal remained the major source up to the Second World War, then lost ground very rapidly in the USA and more slowly in Europe, including the UK. Mr P. V. Youle of ICI summarised the UK position in 1975 in these terms:

Energy and Chemicals from Coal
1945: coal provided 60 per cent of total energy and 75 per cent of the organic chemicals.
1975: coal provided 30 per cent of total energy and 5 per cent of the organic chemicals.

(For perspective it should be added that total UK coal consumption in 1945 was 190 million tons and in 1975, was 118 million.)

The major base chemicals in this category (as noted earlier) are ethylene, propylene, butadiene, benzene, toluene and xylenes. Shell Chemicals estimate that the world total of these materials produced from petroleum shot up from a little over 2 million tonnes in 1950 to almost 70 million in 1973. Three main areas of production are North America, West Europe and—more recently—Japan, outstripping the other areas in its rate of development, and even, despite its later start, approximating to the USA in the weight of these chemicals manufactured per head of the population.

Comments in an economic survey for the UK (by a specialist working party of NEDC) on the large scale activities in basic organic chemicals

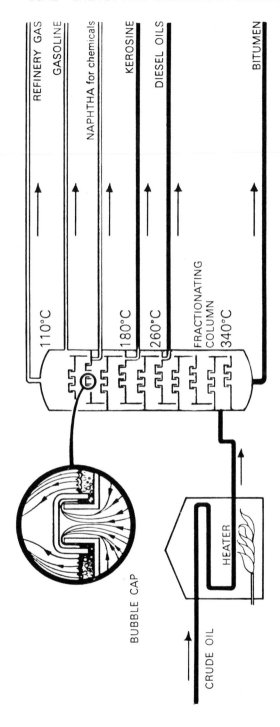

Figure 1. Distillation of oil: the heated crude oil enters the column as vapour and rises through the trays and bubble caps. The enlarged picture shows a bubble cap: it forces the rising vapours to bubble through the liquid already formed on the tray by condensation. This helps to produce a thorough separation of the different fractions. (Shell).

indicate both the concentration and the vertical integration of the industry. The Organic Chemicals Sector Working Party observed

'Much of the sector's output is produced by a handful of firms whose operations are highly capital intensive and are often integrated with the "upstream" oil refinery processes as well as the "downstream" activities such as plastic materials manufacture.'

The broad pattern of the flow of oil through a refinery is shown in Fig. 1 and a more generalised scheme of the relationships to chemicals in Fig. 2.

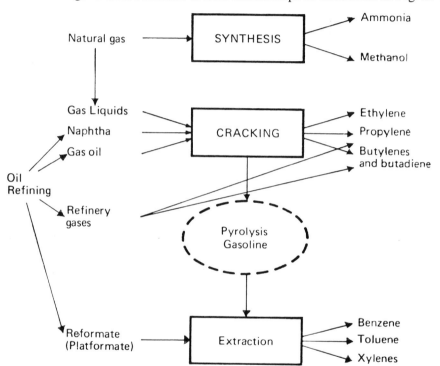

Figure 2. Simple scheme of the relationships between main oil feedstocks and the most important chemical 'building blocks'. (Shell).

In principle at least, distillation, the first separation process for the crude oil is probably familiar. The crude is heated and fed into a tall metal cylinder of a kind that most of us have seen when we pass refineries: it is called a 'fractionating column' since it yields a number of 'fractions' from the complicated mixture flowing into it. In the column (or series of columns) the more volatile parts turn into vapour and rise, becoming cooler. Some of these condense to liquid and are drawn off at various levels thus using the widely varying boiling points as the basis for separation. Bubble cap trays set across the column increase the efficiency of separation. Gases pass off from the top and a residue is drawn off from the bottom. The fractions then go through further stages of separation which may include distilling under vacuum for heavier molecules, or pre-treating and chilling to remove wax from materials destined to make lubricating oils.

After the separation, there may be chemical conversion too.

This is generally based on chemical 'cracking' (Fig. 3), that is breaking down large molecules of the separated hydrocarbons into smaller ones, by heating them at higher temperatures than those of the distillation and

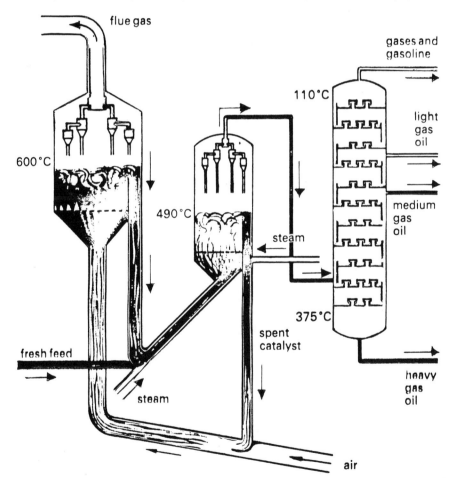

Figure 3. Catalytic cracking: the catalyst powder passes to the reactor (centre) where the cracking process takes place. The cracked vapours then pass to a fractionating column (right) for the separation of lighter and heavier fractions. The used catalyst is blown by an air current back to the regenerator (left) where it is cleaned ready for use again. (Shell).

often with catalysts. Usually known as 'cat cracking', the process takes advantage of the property of certain materials (the catalysts) to speed up the rate of chemical reaction without themselves being consumed—although they may need regenerating, and after a time, replacing. Cracking may also be carried out using hydrogen—also in the presence of a catalyst—for breaking down heavier fractions. Further processes are based on using heat and pressure only, known as thermal cracking. Of very great importance for the chemical industry is steam cracking at high temperatures and low pressures. Then, apart from some fractions being split up in the cracking reactions, others may be chemically modified in other ways, called 'reforming'. Since this too is speeded up by a catalyst, normally containing platinum, the abbreviated description is the fused-together word 'platforming'.

According to the trends in availability and prices of supplies, as well as the demands and prices obtainable for products, the oil refining and petrochemical industries constantly modify their procedures. World-wide these forces have led to an increasing tendency towards cracking heavier oil fractions which tend to be of lower financial value. The refinery product currently used in Europe including the UK as feedstock for making ethylene and propylene and also for aromatics is naphtha (with a boiling point range roughly between 65°C and 170°C). But a higher boiling product called gasoil, otherwise going into the heating market, is likely to be increasingly used as a feedstock. A survey commissioned by the Chemicals EDC to check trends in petrochemicals and look at the new opportunities offered by North Sea oil warned that major flexibility can be built into plants but only at increased costs. Their consultant advised that a major existing petrochemicals complex could best accept a shift to an alternative feedstock as an *extension* to output; it could not change over for all or a major part of its output. However, within narrower limits refineries and petrochemical plants do in fact adjust to changes in feed and market demands, even though on economic grounds they seek to minimise the changes and to ensure that they operate unchanged for periods that are as long as possible.

It is this very flexibility that makes it difficult to construct a diagram that is both comprehensive and not too complex to be readily understood. For chemicals are made from all of these hydrocarbon fractions:

* Naphthas for cracking to ethylene and others in this series of compounds known as the olefins—very chemically active and the most important group of organic base chemicals—but yielding other products too;
* Gas oil, as noted above, cracked to a similar range of reactive chemical building blocks;

* Reformate or platformate. Here the direct distillate has been catalytically 'reformed' to yield benzene and others in a further series of compounds known as 'aromatics', second in importance only to the lower olefins. The principal members, benzene, toluene and the xylenes, are known together by their initials as BTX;
* Refinery gases. These have separated from them the lower olefins and also ethane and others in yet another series of organic compounds known as the paraffins;
* Natural gas. In the UK this is largely methane, the lowest member of the paraffins. Elsewhere it tends to vary more widely in composition, containing ethane and higher members of the paraffin series, but also hydrogen sulphide, carbon dioxide and other constituents. Notably in the USA ethane, propane and butane are cracked to lower olefins;
* Condensate from the gases associated with oil may itself contain a range of materials from ethane to naphthas, treated similarly to those from refining;
* Petroleum wax may also be cracked, though in this case it is 'higher olefins', that is with more carbon atoms in the molecule, which are obtained;
* Fuel oil may be cracked forming a synthesis gas providing scope for additional routes of chemical building.

Further complications in striving to follow the pattern of procedures arise from the fact that chemicals may arise partly (as they did wholly at the start of the petrochemical industry) as by-products when the primary aim is making something else. So, for example largely in the USA, propylene, second member of the olefin series, is formed mainly as a by-product of making gasoline by catalytic cracking. Furthermore, the next olefin—in this case a group of isomers known as the butylenes all having the same number of carbon atoms, that is four, but different chemical structures— may be sent for blending into gasolines or for chemical synthesis.

Despite these complications, a highly generalised scheme has been suggested by Shell Chemicals, and has the appearance shown on page 24.

In Britain the government have indicated that they are likely to try and encourage the development of petrochemicals from what are called 'heavy natural gases'—ethane and possibly higher paraffin gases—associated with North Sea oilfields. In December 1976, the Department of Energy announced setting up studies on a possible gas gathering line. If this is built, and also the associated chemical plant, it will of course mean that the routes shown as being used 'almost entirely in the USA' will also be introduced in the UK.

For readers who would like to understand a little more about the terms

C

Lower olefins & diolefins—petroleum sources of manufacture

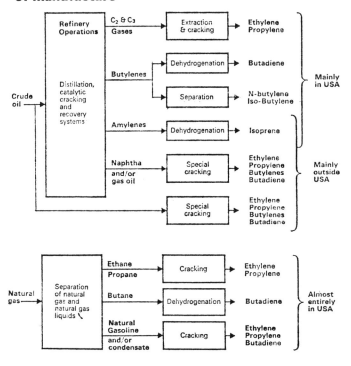

Aromatics—main petroleum sources of manufacture

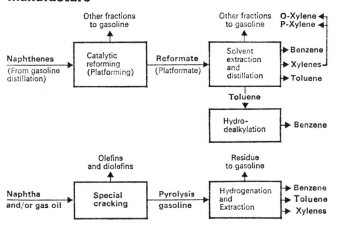

Figure 4. (Shell)

and symbols used in the diagrams of Fig. 4, I have added a technical appendix at the end of this chapter.

Some familiar derivatives

It is evident that all the chemical and process ingenuity is prompted by ultimate consumer needs, although these may seem very remote at the stage of say cracking naphtha and separating the complex mixture that results. For in addition to gasoline and fuel oil that a consumer would immediately recognise as usable, there may also be methane, hydrogen, ethane, ethylene, propylene, propane, aromatics, butadiene and others. What is their significance in terms of end-products?

The name ethylene immediately suggests its polymer (the giant molecules formed by large numbers of the single ones combining together), polyethylene. But the same olefin is also the starting point for further plastics, PVC and polystyrene and for synthetic rubbers; from it too is made antifreeze, polyester fibres and synthetic ethyl alcohol with a wide range of solvent applications as well as being itself a starting point for syntheses and a special fuel for rockets.

Propylene too fairly obviously transforms into the familiar polypropylene, then less expectedly serves to form such materials as acrylic fibres, polyurethane resins, nylon. In the case of the next higher olefin however, the butylene compounds, and butadiene which also has four carbon atoms, it is synthetic rubbers rather than other forms of plastic that are characteristic end-products.

The aromatics—benzene, toluene and xylenes—have traditionally been obtained from coal tar. In a later chapter we shall examine how much is still contributed in this way and how it arises. At this stage let us note that about nine-tenths of both world and of UK supplies originate from petroleum sources, during platforming of a 'cut' or fraction to improve the quality and yield of gasoline and also from cracking naphtha. Benzene too is used for making nylon—and it should be mentioned that nylon is not a single material of specified composition but a family of chemically related synthetics containing a characteristic group known as polyamides. For the familiar cheap polystyrene plastics too, benzene is essential and a further important outlet is in making detergents.

Varying the route

In the above scheme I mentioned in passing that ethanol (made from ethylene) is itself the starting point for further syntheses. One of its

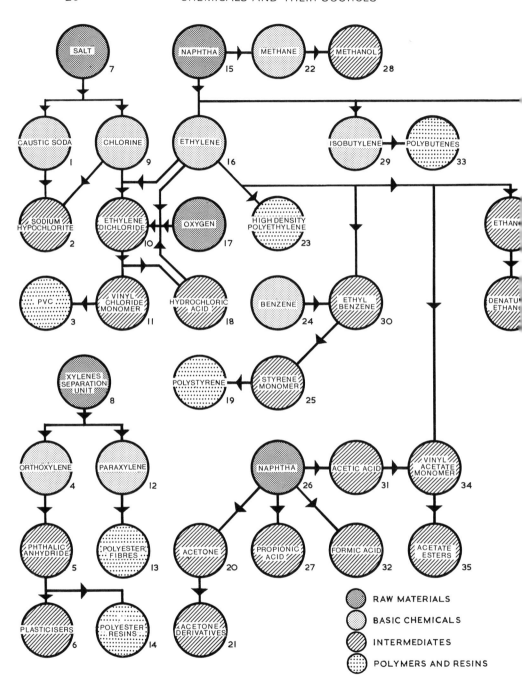

Figure 5. Major chemicals produced in the UK. (BP)

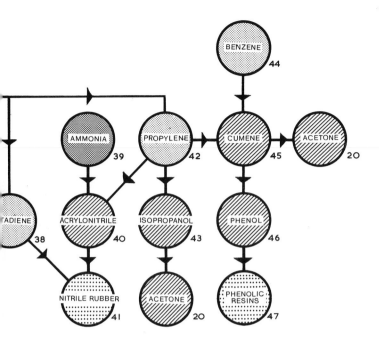

1 Caustic soda
Basic chemical for petroleum refining, pulp and paper manufacture, textile processing, rubber reclamation, etching and electroplating.

2 Sodium Hypochlorite
Intermediate forming the basis of industrial and domestic bleaching agents.

3 PVC
Polymer in rigid or flexible form for cable and wire coatings, insulations, pipes, hoses, flooring, packaging, building products, records, toys, travel goods, rainwear goods, footwear etc.

4 Orthoxylene
Basic chemical for phthalic anhydride.

5 Phthalic anhydride
Basic chemical for plasticisers, polyester resins and alkyd resins.

6 Plasticisers
Additives for imparting flexibility to PVC.

7 Salt
Raw material for manufacture of caustic soda and chlorine.

8 Xylenes separation unit
Raw material for orthoxylene and paraxylene.

9 Chlorine
Basic chemical for ethylene dichloride and vinyl chloride monomer. Also used for water purification, pesticides, disinfectants and bleaches.

10 Ethylene dichloride
Intermediate for vinyl chloride monomer and solvents.

11 Vinyl chloride monomer
Intermediate for manufacture of PVC.

12 Paraxylene
Basic chemical for polyester fibres.

13 Polyester fibres
Polymer—one of the major types of synthetic fibre, e.g. Terylene.

14 Polyester resins
Polymer, reinforced with glass fibre used for manufacture of vehicle bodies, vehicle components, boat hulls and components, translucent roofing, building materials and finishes etc.

15 Naphtha
Liquid petroleum hydrocarbon used as feedstock for ethylene, propylene, butadiene, isobutylene and acetone.

16 Ethylene
Versatile basic chemical for manufacture of polyethylene, vinyl acetate monomer, ethanol, styrene monomer, polystyrene, ethyl benzene etc.

17 Oxygen
Raw material for manufacture of ethylene dichloride used in conjunction with hydrochloric acid.

18 Hydrochloric acid
Intermediate used in conjunction with oxygen for manufacture of ethylene dichloride by oxyhydrochlorination.

cont'd. overleaf

cont'd

19 Polystyrene
Polymer for manufacture of radio and TV cabinets, cosmetic packs, packaging, refrigerator linings, film cassette cartridges, toys etc.

20 Acetone
Intermediate for chemical production and solvent use.

21 Acetone derivatives
Intermediates for methyl methacrylate, solvents, chemical intermediates etc.

22 Methane
Basic chemical for methanol.

23 High density polyethylene
Polymer for household, industrial and medical articles, packaging, mouldings, containers, undersea cable sheathing, synthetic yarn, film and sheet, large blow mouldings, milk bottles, tissue film.

24 Benzene
Basic chemical for cumene in conjunction with propylene and ethyl benzene in conjunction with ethylene.

25 Styrene monomer
Intermediate for manufacture of polystyrene and styrene-butadiene rubbers.

26 Naphtha
Liquid petroleum hydrocarbon used as feedstock for ethylene, propylene, butadiene, isobutylene and acetone.

27 Propionic acid
Intermediate used for herbicides and plastics; basis of Propcorn moist grain preservative and Add-H mould inhibitor for hay.

28 Methanol
Intermediate for pharmaceuticals and fine chemicals. Used in the manufacture of formaldehyde, solvent for surface coatings, polishes, stains and lacquers,

also for insecticidal and fungicidal sprays, and as an alcohol denaturant.

29 Isobutylene
Basic chemical for manufacture of polybutenes and butyl rubber.

30 Ethyl benzene
Intermediate for manufacture of styrene monomer.

31 Acetic acid
Intermediate for manufacture of cellulose acetate for man-made fibres and plastics, vinyl acetate, acetate ester solvents, food and tobacco preservatives, herbicides, pharmaceuticals processing aids.

32 Formic acid
Intermediate used in textile and leather processing; the basis of Add-F silage additive.

33 Polybutenes
Polymers for use in oil additives, adhesives, sealing agents etc.

34 Vinyl acetate monomer
Intermediate for manufacture of emulsion paints, adhesives, plastics and textiles.

35 Acetate esters
Intermediate in the formulation of surface coatings, adhesives etc.

36 Ethanol
Intermediate for manufacture of industrial methylated spirits, adhesives, pharmaceuticals, fine chemicals and fine essences, insecticides and solvents for spirit lacquers.

37 Denatured ethanol
Intermediate used in formulation of cosmetics, perfumery, detergents, printing inks etc.

38 Butadiene
Basic chemical for manufacture of synthetic rubbers including

styrene-butadiene and polybutadiene rubbers.

39 Ammonia
Raw material used in conjunction with propylene fo manufacture of acrylonitrile.

40 Acrylonitrile
Used for synthetic fibres in conjunction with butadiene fo nitrile rubbers; in conjunction with butadiene and styrene for ABS plastics.

41 Nitrile rubber
Polymer for telephone cables, hoses, oil seals, brake linings, adhesives, footwear, non-wove fabrics etc.

42 Propylene
Versatile basic chemical for manufacture of acrylonitrile, cumene, isopropanol, acetone etc.

43 Isopropanol
Intermediate for acetone; used as a solvent in synthetic lacquer and cosmetics.

44 Benzene
Basic chemical for cumene in conjunction with propylene and ethyl benzene in conjunction with ethylene.

45 Cumene
Intermediate for phenol and acetone.

46 Phenol
Intermediate for the manufacture of phenolic and other resins, insecticides, herbicides, synthetic fibres, lub oil additives, dyestuffs, pharmaceuticals and fine chemicals.

47 Phenolic resins
Resins used for the manufactu of electrical components, closures, appliance parts, handles, telephone equipment, laminates, chipboard etc.

important derivatives is acetic acid. But BP Chemicals have developed a process for making this familiar sounding condiment much more directly—by adding oxygen to naphtha so cutting out several steps. The importance of this is that acetic acid, far from being just a minor flavouring for food, is a major industrial chemical consumed at a rate approaching 200,000 tonnes a year. Most of it in the UK is made in this way and the interaction with a modified processing scheme is shown in Fig. 5. About a third goes into making cellulose acetate for synthetic fibres, plastics and packaging while another large slice is applied to producing vinyl acetate for emulsion paints and adhesives; smaller proportions help to make pharmaceuticals and of course, some does go as we all know for culinary use. The process for direct manufacture of acetic acid from petrochemical feedstock is operated under licence in other countries, the feed being chosen according to availability and local cost conditions.

Prospects for supplies

This then in outline is a broad picture of the way in which we now obtain ammonia and the range of organic chemicals on which we depend for our style of life. And on a time-scale of some two to three decades, it is liable to be threatened because supplies of both the oil and the gas are estimated by the relevant experts within and without those industries as likely to be declining. It is for these (secondary) reasons, backing the primary ones of ensuring energy supplies, that the oil and gas companies are now buying into the coal industries in many countries and, in respect of chemicals, are looking first at ways of supplementing their sources; later they will inevitably have to replace them.

What then is the chemical character of coal and which processes can transform it competitively into the chemicals, as well as the energy sources, that we shall need to provide much more extensively from it in the future?

Technical Appendix

The term 'C$_2$ gases' refers to ethane, second member of the paraffin series, with the formula C$_2$H$_6$, and also to ethylene—lowest member of the olefin series—having the formula C$_2$H$_4$. (see Figs. 6 & 7). Correspondingly, 'C$_3$ gases' means the next higher member of each series—the fully saturated paraffin, propane, C$_3$H$_8$, and the unsaturated olefin propylene C$_3$H$_6$. Butylenes are the next olefin compounds in the series with formula C$_4$H$_8$ and those immediately above them with formula C$_5$H$_{10}$ are the amylenes

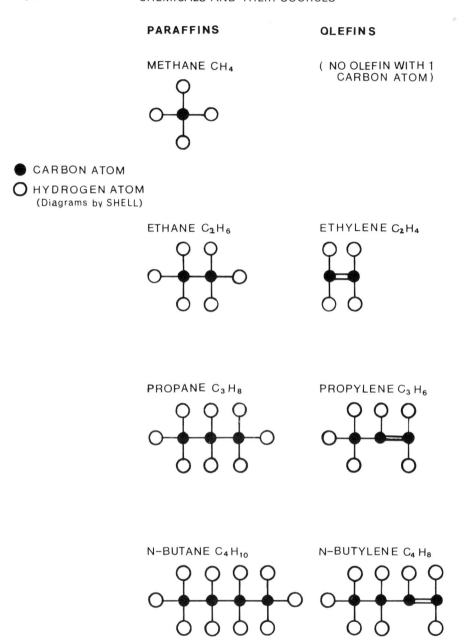

Figure 6. Basic hydrocarbon structures: Paraffins are simple saturated hydrocarbons. The olefins are closely related but are unsaturated and contain a reactive double bond.

AROMATICS

BENZENE C_6H_6

TOLUENE $C_6H_5 \cdot CH_3$

XYLENES $C_6H_4(CH_3)_2$

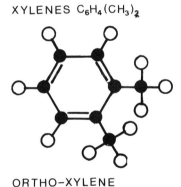

ORTHO–XYLENE

META–XYLENE

● CARBON ATOM
○ HYDROGEN ATOM
(Diagrams by SHELL)

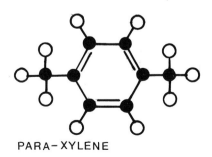

PARA–XYLENE

Fig. 7 Basic aromatics structures.

better called 'pentenes'. When the butylenes are formed among the products of cracking naphtha—which is generally the case in European practice—the C_4 stream also contains butadiene; this, as the latter part of its name indicates, contains two chemically unsaturated bonds. Butylene derived from catalytic crackers in petroleum refineries (the U.S. situation) contains no butadiene and this important intermediate then has to be obtained as shown in Fig. 6 by removing some of the hydrogen from the butylenes.

Turning to the aromatics, the main chemical point to note is that their carbon atoms form a closed ring structure unlike the straight or branched chains of the simple compounds mentioned above. Sometimes they are combined with the olefins but they may be used for forming other derivatives.

Among the other terms used in the diagram, the word 'naphthenes' may not be very familiar. These are again compounds where the carbon atoms are in a ring but larger proportions of hydrogen are combined with them so that they have markedly different properties; they are known as cyclo-paraffins or hydro-aromatics.

Finally the word 'de-alkylation' may justify some explanation. Toluene (formula C_7H_8) consists of benzene, the first compound of the aromatic series C_6H_6, in which in effect one of the combined hydrogens has been displaced by a combined group CH_3 known as the 'methyl radical.' Thus toluene is also known as methylbenzene. The general term for this series of radicals is the 'alkyl' radicals. So converting toluene to benzene is chemically a matter of combining with it hydrogen and removing from it an alkyl; the single word that says all this is hydro-dealkylation, or simply de-alkylation.

Further Reading

'UK Chemicals 1975–1985.' Report of the Chemicals Economic Development Council, NEDO, 1976.

'Chemicals Information Handbook, 1975–6.' Shell International Chemical Co. Ltd, London (and other Shell chemical booklets).

'Chemical Technology.' Professor F. A. Henglein, Pergamon Press, London, 1969 (translated from the German). This is a detailed and extensive survey.

'Profile of BP Chemicals.' BP, Dec, 1976 (and other BP chemical booklets).

Chapter 3

Chemical character of coal

What is the nature of coal? Nowadays a far less familiar sight in the home than it was a generation ago, it would nevertheless be generally recognised as a blackish solid that can be burnt. It is not a single material of one chemical composition but a range of natural solids rich in carbon. The complete range is broadly taken to extend from peat—still clearly a breakdown product of vegetable matter—via brown coal, lignite, and bituminous coal to anthracite.

Coal is the fossilised product of decomposition of abundant tropical forest growths under marshy conditions. Within it is a proportion of material that will not burn—largely the residue of silt deposited between and on the rotting vegetable matter. It is this incombustible part that yields ash when coal is burnt. In the swamps, plants would grow thickly over long periods of time and their debris would accumulate. Then the area subsided and became flooded, putting an end to growth but also resulting in the vegetable remains being covered with sand and silt. Later the area was lifted by further earth movements and another generation of swamp forest began to grow. The process was repeated many times during periods measured in millions of years.

Over these very long periods—rather difficult to envisage when most of us think of the times since the start of written human history (only thousands of years) as incredibly distant—the plant components were partly decomposed and compressed. First, micro-organisms needing oxygen from the air ('aerobic'), then others that live without air ('anaerobic') broke down much of the plant protoplasm, cellulose and other parts, though waxes and resins proved more resistant. Further weight of sediments deposited over the tops of these layers increased the pressure and they became slightly warmed. Understandably, the period when plants were being broken down is called the 'biochemical stage'; the later period of compression and mild heating is the 'geochemical stage'.

Artist's impression of a swamp forest of the kind that, millions of years later, was transformed into coal. (NCB).

The name given to the whole sequence of changes is 'coalification'—and how far it has gone is known as the 'rank' of the coal. The whole process is generally more advanced in the older seams, though coals are found of differing rank even at the same geological age since earth movements, pressures and temperatures have all varied enormously. These variations in effects are quite marked, sometimes even within limited areas; consequently a coal seam may change appreciably in thickness and in many cases in rank as it is traced across a coalfield.

The successive layers of decaying plant residues in due course became a series of layers (seams) of coal and the intervening layers, also compressed and hardened, became rock strata such as sandstone—born of sand—and shale, a stone resulting from what had been clay.

Are there swamps of this kind around today? Some researchers have suggested that the Dismal Swamp of Virginia and North Carolina, which is gradually being flooded by Lake Drummond, is an area of that type now undergoing active subsidence. There are also debates among coal scientists on whether all or particular coal seams arose from plants growing on the same sites or from plant debris that had drifted into deep enclosed basins. But there is general agreement that coal originates from plants.

Composition

What then is the essential chemical composition? Since we are not dealing with a single pure chemical, the answer can be given only as ranges. And the further complication of having a varying amount of mineral matter as impurity means that composition is most usefully considered when the figures are stated on the basis of the organic matter only, free of the associated mineral matter and moisture. In the table below the first column shows the moisture of coal *as found*, but all the others give the percentage of the respective chemical elements or properties on this corrected basis for the pure coal.

The term 'humic' refers to the main range of common banded coal. (There are others based on different plant origins and differences in conditions during coalification.) As applied in the analysis of coal, the expression 'volatile matter' refers to the percentage loss in weight when the coal is heated in the absence of air under standard conditions. Although this is apparently a purely practical test—what is called an 'empirical' one—it is of great value, since it is fast, it can be used for classifying coal and it has proved to give results closely related to those obtained by more fundamental methods, such as true chemical analysis. Calorific value is the measure of the heat obtained under very closely standardised conditions by burning the coal. This too is related to the carbon and hydrogen content.

Composition of Main Types of Humic Coals.

Type of coal	Moisture as found	Carbon	Hydrogen	Oxygen	Nitrogen	Volatile matter	Calorific value Btu/lb
		all on dry, mineral-matter-free basis					
	%	%	%	%	%	%	%
Peat	90–70	45–60	6·8–3·5	45–20	0·75–3	75–45	7,500–9,600
Brown coals and lignites	50–30	60–75	5·5–4·5	35–17	0·75–2	60–45	12,000–13,000
Bituminous coals	20–1	75–92	5·6–4·0	20–3	0·75–2	50–11	12,600–16,000
Anthracites	1·5–3·5	92–95	4·0–2·9	3–2	0·5–2	10–3·5	16,000–15,400

(slightly re-arranged from Dryden, article on 'Coal' in Kirk–Othmer Encyclopedia of Chemical Technology, 1965)

The more the original plant remains have been altered by heat and pressure after they have been buried, the higher the rank of the coal. The percentage of carbon in the pure organic matter is taken as the measure of rank but the table shows how other properties are also related. There is always moisture associated with coal; as rank increases this moisture falls from the very high proportions in the soggy, partly-decomposed plant debris of peat to the one or two per cent associated with the highest rank coals and the anthracites. Across that same range, the hydrogen contents of coals also fall and so does the oxygen, but much more sharply. And rank is reflected too in the values obtained for the volatile matter.

One of the pioneering giants of coal science (C. A. Seyler) drew up a composite chart many decades ago relating these properties. Later others added more properties on to the same chart—hammering home the point that many aspects of the behaviour of different coals in their chemistry, physics, burning and liability to make coherent coke when they are heated, are all related and closely connected with rank.

A simplified from of this chart is shown as Fig. 1. You can readily understand it if you take it by easy stages. The main frame is made up of the scales for carbon and hydrogen respectively; within it the tilted frame is based on values for volatile matter and calorific value. A further set of curved lines marked with numbers refer to a British Standard test for how much the coal swells up on heating—one of the relevant properties in making coke.

On this background of axes for various properties is drawn a band enclosed by the two lines, with a centre line added. Most bright coals fall within that band. Outside it are other coals, generally of less significance. Among these are some coals called 'cannel' (said to be a perversion of the word 'candle' since they burn readily with a long smoky flame) which are dull, hard and have high volatile matter. They are thought to originate from different plant remains and are found near normal coals in all UK coalfields but especially in Scotland and Lancashire. Coals found in newly-developing coalfields in South America are also reported to fall outside the marked band.

According to Dr W. A. Read of the Institute of Geological Sciences the forming of coal started much earlier in Central Scotland and the extreme north-east of England than elsewhere in Britain. The earliest coals of workable thickness lie within groups of rock layers called respectively the Caliciferous Sandstone Measures of Scotland and Scremerston Coal Group of Northumberland. Both are thought to be about 340 million years old. The geological division known as 'Carboniferous' (345 to 270 million years ago) is, as the name suggests, the chief coal-forming period, particularly in the Northern Hemisphere with brackish swamps over much of Europe and North America. However there are also important

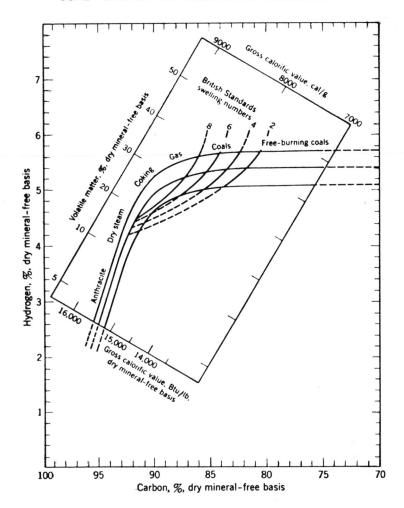

Figure 1. Simplified form of Seyler's coal classification chart. (after Dryden, article on 'Coal' in Kirk-Othmer encyclopaedia of chemical technology 1965. John Wiley & Sons Inc.).

deposits dating from the Cretaceous (from 135 to 70 million years ago) and there are lower rank coals thought to have been formed in this period. All peat is reckoned to be geologically recent, that is the product of less than one million years.

Coal has also been examined as a rock by the group of techniques generally described as petrography involving observations in the field, on hand-specimens and in thin sections. Examined by transmitted light in thin sections under the microscope it has been found to have a complex fine structure, and also surprisingly displays a variety of colours. Orange to dark red bands often show the original cell structure of wood; others consist of finely divided plant remains and there are yellow translucent forms that are identified as resins, algae, the outer layers of spores and pollen (called exines) and the surface layers (cuticles) of plants. Polished surfaces have also been studied under the microscope by reflected light, showing differences in the microscopic layers first noted some sixty years ago. By analogy with the fine minerals in other rocks, 'macerals' were identified and named for the fine structure of coal. Before going into this degree of fine detail, it is possible to distinguish by eye these four components: vitrain—the shining black part with a glassy lustre; it occurs in thin bands but does not show banding within itself and is thought to represent single pieces of plant material such as bark; durain—dull and hard sometimes flecked with thin streaks of bright coal containing masses of resin and plant cuticle; clarain—bright coal but not truly a different component since it can readily be seen to be fine alternating layers of durain and vitrain; fusain—the soft, friable, fibrous part that dirties your hands; it is found in patches. Once thought to be a true charcoal, the residue of natural forest fires, it is interpreted as probably formed from woody parts of plants during dry conditions.

Within vitrain and durain are found a number of macerals showing finer structure. Yet all this concerns only the complications of the organic coal substance. To it are added in practice the inorganic material, both finely dispersed and in layers, arising from disintegration of other rocks when the plant debris was being laid down and at the early stages of conversion to peat. Some can have been introduced later from water percolating through the seams. Clay is the most common but a large number of other materials are also found in coal such as quartz, feldspar, pyrites and others. Coal mineral matter and its ash have been carefully analysed particularly with an eye to finding valuable trace metals. It has proved to have a wide range of metallic elements. There is a high proportion of aluminium and there have been proposals to develop processes for extracting this useful metal—though most assessments conclude that this is not competitive. Germanium has been extracted for use in transistors.

The metals present only in traces are presumed to have been combined mainly in the plant material; among them are vanadium, titanium, gold, silver, uranium, platinum and other precious or esoteric sounding metals, but concentrations are low.

This inorganic mineral matter, present in all coals, causes complications in chemical analysis and practical testing. The most simple way of determining how much is present is to heat the coal under standardised conditions and weigh the proportion of residue—the ash. However this is not the same as the original mineral matter, since the minerals are also chemically changed during the ignition. Combined water is driven off from silicates, sulphur is oxidised—some coming off as sulphur dioxide, some being 'fixed' in the ash as sulphate—and there are other changes. Considerable research has gone into finding arithmetical formulae that make it possible to calculate the original proportion of mineral matter in a coal 'as-received' from the figures for moisture, ash, sulphur and carbonates determined by analysis.

This is not an academic exercise. For it is only by first physically reducing the quantity of the mineral matter before testing (as far as is practicable), and then allowing for what remains intimately mixed with the organic coal, that it is possible to derive values for properties of the true coal substance itself. Consequently, although coal as-received may be quickly characterised by determining and referring to its 'ash content', valid analyses for other properties must be related to the coal, dry and free of mineral matter. This is called the 'dry, mineral-matter-free basis' or dmmf. Earlier, analyses were converted by using the figure for the (moisture and) ash itself without conversion back to mineral matter; this was called the 'dry, ash-free basis' or daf and older analyses are on this less accurate basis.

Classifying Coal

For classifying the rank of the coal substance of a coal, the NCB use a system based on the percentage of volatile matter and the type of coke produced in a standard test (the Gray-King coke type). For these tests coals of over 10 per cent ash must be 'cleaned'. They are treated in a liquid where the mineral matter sinks because of its higher specific gravity and the coal floats.The process is arranged to give a maximum yield of coal with ash of 10 per cent or less. Cleaning is essential because large amounts of mineral matter in the coal seriously depress the 'caking properties' as determined in the standard test, apart from the arithmetical effects on the figures of the analyses discussed above. Even after this cleaning, the volatile matter must be corrected to the dmmf basis. In turn this means that the coal must be analysed for at least moisture, ash, total sulphur and carbonates in addition to the volatile matter and the coke type. When these are all completed, the coal can be classified using the diagram shown as Fig. 2 or a corresponding table.

As the diagram shows, when the volatile matter is less than 19·5 per cent the value found directly determines the class of the coal. The dotted lines define limits of coke type (from the Gray-King test) found in

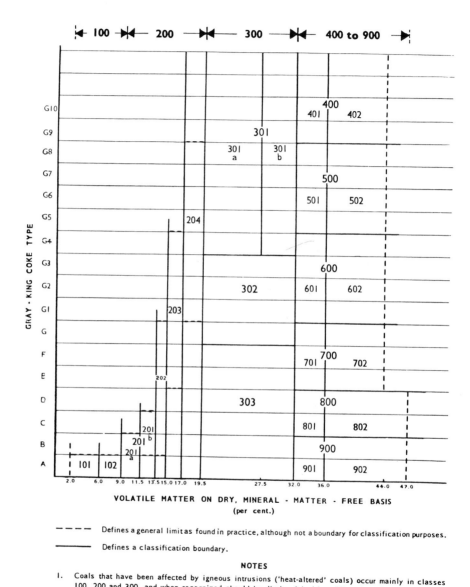

VOLATILE MATTER ON DRY, MINERAL - MATTER - FREE BASIS
(per cent.)

- - - -   Defines a general limit as found in practice, although not a boundary for classification purposes.

————   Defines a classification boundary.

NOTES

1. Coals that have been affected by igneous intrusions ('heat-altered' coals) occur mainly in classes 100, 200 and 300, and when recognized should be distinguished by adding the suffix H to the coal rank code, e.g. 102H, 201bH.

2. Coals that have been oxidized by weathering may occur in any class, and when recognized should be distinguished by adding the suffix W to the coal rank code, e.g. 801W.

Figure 2. Coal classification system used by NCB. (Revision of 1964).

practice but are not class boundary lines. From 19·6 per cent volatile matter, and above. the coals are classified by using both the figure for volatile matter and also the coke type. The term 'igneous intrusions' mentioned in the note below the diagram is used in geology (a little misleadingly) to mean the condition when molten rocks generated deep down in the earth were pushed up into other layers—in this case into or near to coal seams so that they were much more strongly heated than the others. Where this has happened, the nearby coals tend to have properties that do not correspond with each other in exactly the same way as for 'normal' coals.

All the major coal countries have felt the need for a system for dividing coal into classes by rank. The American standard specification for this purpose, for example, opens with the statement that it is classifying coals 'according to their degree of metamorphosis, or progressive alteration, in the natural series from lignite to anthracite.' Their method is to use the value for 'fixed carbon'—the residue left after volatile matter has been driven off—for higher rank coal, which is in practice the same principle as described earlier for the UK; the lower rank coals are then classed according to calorific value, a different criterion but as we saw earlier related to the coke-making property.

A much more complex system has proved to be necessary for setting up an international classification. 'Necessary' it was, because of the needs of the international coal trade; and it had to be more complex since it was to deal with a far wider range of coals than in any single national system. Consequently the classification of the Economic Commission for Europe (ECE) gives a coal a three digit number. Of these the first is defined by volatile matter, the second by the caking properties shown when they are heated rapidly and the third by the coke-making behaviour when they are heated slowly, as in the Gray-King test I mentioned above. For the first digit, calorific value is used as a basis for classifying the coals of volatile matter above 33 per cent.

Before this was evolved there were many exercises by international groups of coal scientists comparing the various national approaches. And when the early drafts of the ECE scheme were proposed there were further examinations—I took part in some of them—to check that it would sensibly separate coals that were appreciably different in use and group together those that were reasonably alike. A further international co-operative effort was also needed to verify that all known coals would actually be capable of being classified using the system. It was only after many years of discussions and checks of these kinds that ECE published the system in 1956.

How the various systems compare can be seen from the composite table forming Fig. 3.

Bases of various coal classification systems

Coal classification system	Principal parameter	Subsidiary parameter(s)
American Society for Testing and Materials (ASTM)	Lower-rank coals: CV (mmmf) Higher-rank coals: FC (dmmf) or VM (dmmf)	Agglomerating character Agglomerating character
International (ECE)	Soft coals: moisture (af) Lower-rank hard coals: CV (maf) Higher-rank hard coals: VM (daf)	Tar yield (daf) Caking and coking properties Caking and coking properties
National Coal Board (UK, NCB)	VM (dmmf)	Gray–King coke type
Seyler	C in vitrain (dmmf)	H (dmmf)
Australian	Lower-rank hard coals: CV (daf) Higher-rank hard coals: VM (dmmf)	Crucible swelling number, Gray–King coke type, and ash (d) Crucible swelling number, Gray–King coke type, and ash (d)
Indian	VM (dmmf)	CV (dmmf) and moisture (mmf)

CV — calorific value; FC — fixed carbon; VM — volatile matter
d — dry basis; af — ash-free basis; maf — moist, ash-free basis; daf — dry, ash-free basis; mmf — mineral-matter-free basis; mmmf — moist, mineral-matter-free basis; dmmf — dry, mineral-matter-free basis

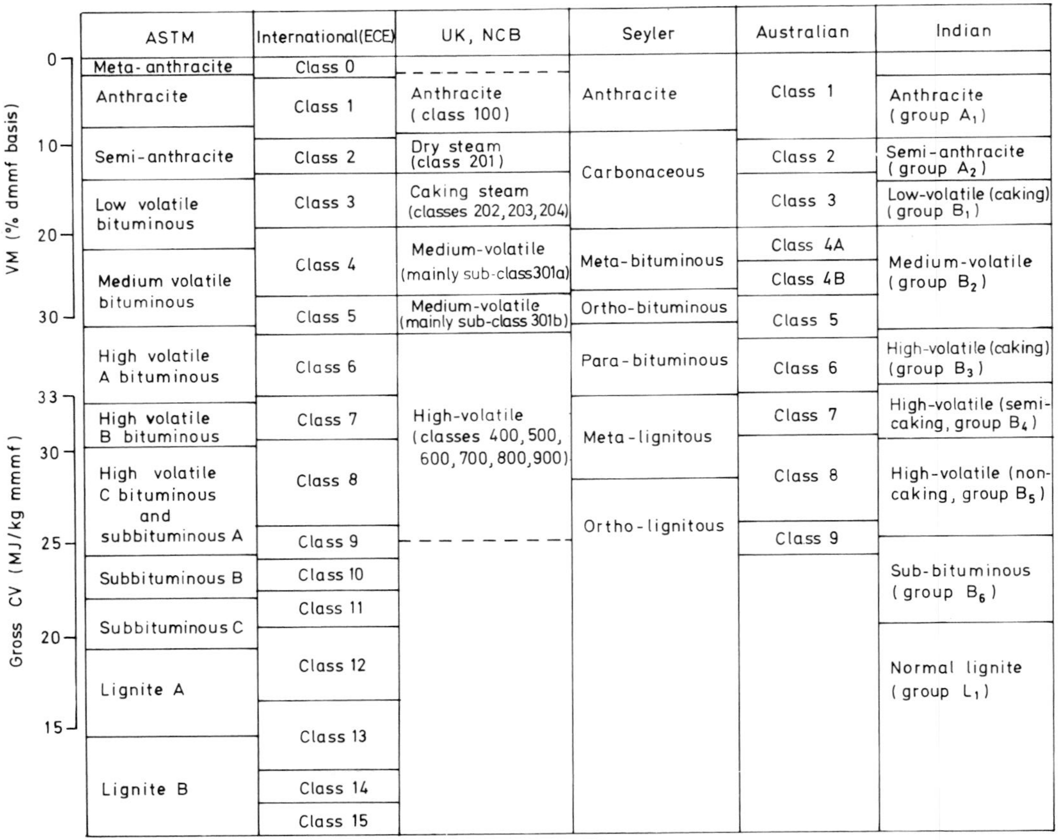

Figure 3. Bases of various coal classification systems, showing the correspondence between classification schemes.

ASTM
- Meta-anthracite
- Anthracite
- Semi-anthracite
- Low volatile bituminous
- Medium volatile bituminous
- High volatile A bituminous
- High volatile B bituminous
- High volatile C bituminous and subbituminous A
- Subbituminous B
- Subbituminous C
- Lignite A
- Lignite B

International (ECE)
- Class 0
- Class 1
- Class 2
- Class 3
- Class 4
- Class 5
- Class 6
- Class 7
- Class 8
- Class 9
- Class 10
- Class 11
- Class 12
- Class 13
- Class 14
- Class 15

UK, NCB
- Anthracite (class 100)
- Dry steam (class 201)
- Caking steam (classes 202, 203, 204)
- Medium-volatile (mainly sub-class 301a)
- Medium-volatile (mainly sub-class 301b)
- High-volatile (classes 400, 500, 600, 700, 800, 900)

Seyler
- Anthracite
- Carbonaceous
- Meta-bituminous
- Ortho-bituminous
- Para-bituminous
- Meta-lignitous
- Ortho-lignitous

Australian
- Class 1
- Class 2
- Class 3
- Class 4A
- Class 4B
- Class 5
- Class 6
- Class 7
- Class 8
- Class 9

Indian
- Anthracite (group A_1)
- Semi-anthracite (group A_2)
- Low-volatile (caking) (group B_1)
- Medium-volatile (group B_2)
- High-volatile (caking) (group B_3)
- High-volatile (semi-caking, group B_4)
- High-volatile (non-caking, group B_5)
- Sub-bituminous (group B_6)
- Normal lignite (group L_1)

VM (% dmmf basis): 0 — 10 — 20 — 30

Gross CV (MJ/kg mmmf): 33 — 30 — 25 — 20 — 15

Figure 3. Bases of various coal classification systems. (From P. L. Rumsby in 'The Efficient Use of Energy' edited by I. G. C. Dryden. IPC).

As rank increases

What are the changes during coalification? It is evident from what I have quoted on approaches to putting coal into classes, that as rank increases coals tend to give ever stronger cokes when tested by standard procedures, down to volatile matter levels of roughly 30 to 20 per cent. Then, with increasing coalification, the coking property weakens until the anthracites with 9 per cent or less of volatile matter are found to yield only a non-coherent powder when heated away from air. The general trends in chemical composition have been mentioned above but a wide armoury of scientific weapons has been applied in striving to gain deeper insights into their significance. Broadly the changes are found to relate to the proportion of their carbon content that is combined in the form called 'aromatic'. This term was briefly mentioned in the previous chapter. It is used to describe compounds where the carbon is combined into a particular kind of 6-membered ring—and the basic exemplar is benzene. Organic compounds of the benzene series, toluene, xylenes, mesitylene and others are all called 'aromatic'. So are compounds where the structure is that of benzene rings 'condensed' or 'fused' together. Examples are naphthalene, anthracene, phenanthrene.

There must be some basis originally for using the name 'aromatic' in this way, some presumed link between a benzene compound and the term used in its conventional sense of indicating spicy or fragrant. But certainly today in chemical usage it is the adjective applied to a major branch of organic chemistry, particularly in differentiating the compounds with one or more of these stable 6-member carbon rings of the benzene type from others—both the open chain type and those with carbon rings not of the benzene type broadly known as naphthenes. What has this to do with coal?

The molecules of bright coal are considered to have clusters of between one and four fused benzene rings linked together. As coalification proceeds these linkages themselves are partly transformed into aromatic type of bonds. And so coalification, chemically, is connected with chemical condensation. Standard methods for showing up these fused together structures show that increase of rank means ever more of this chemical packing together into these condensed structures.

One further point with some relevance to later discussions is that sulphur in coal is partly in the mineral matter—this can be removed by processes that clean the coal of mineral matter—but also partly combined in the organic coal substance itself. The latter clearly stays with the coal during cleaning. In Britain and very often elsewhere, total sulphur amounts to less than $1\frac{1}{2}$ per cent by weight of the coal as mined from the seam.

This then is the complex 'chemical' from which we are considering the possibilities of making a wider range of commercially wanted conventional chemicals as well as continuing to extract some energy values. It is a range of materials of varying compositions, any one of which itself shows heterogeneities within itself—while almost all of them are contaminated by mineral matter that dilutes the coal substance and hinders its chemical processing.

To what extent is it already a source of useful chemicals? Are existing processes capable of further development, or are entirely new approaches needed? These are among the questions to be examined in following chapters.

Further Reading

'Coal'. Article in Kirk-Othmer Encyclopaedia of Chemical Technology 1965 2nd Edition Vol. 5, pp 606–678 by I. G. C. Dryden. John Wiley & Sons Inc.

'The Coal Classification System used by the National Coal Board' (revision of 1964). National Coal Board.

'Standard Specification for Classification of Coals by Rank' ASTM Designation D 388–66 (reapproved 1972). American Society for Testing Materials.

British Standard 1016. 'Coal and Coke, Analysis and Testing'. In 16 parts.

'Parameters of the NCB Classification as a Criterion of the Rank of Coal' by I. Berkovitch. Coke & Gas p. 106 March 1958.

'The Efficient Use of Energy', edited by I. G. C. Dryden. IPC, 1975. (In this, the composite table is in a section by P. L. Rumsby, Appendix A6)

Chapter 4

Chemicals from coal—past and present

While eye of newt and toe of frog, adder's fork and blind-worm's sting—and the rest of the witches' brew—were exaggerated figments as ingredients of a potent potion, it is nevertheless true that animal materials have long been used as sources of chemicals and drugs. Plants, too, were gathered so that their extracts could be applied as chemicals, notably for use as dyes or pharmaceuticals. But from the middle of the nineteenth century it was coal that became the main source of organic chemicals and of ammonia throughout the industrialised world. The chemicals were produced as by-products when coal was heated in the absence of air to yield gas—a process known as carbonisation, or in its modern developments as pyrolysis.

Naturally the early experimenters discovered that tar, a watery ammoniacal liquor and coke were formed at the same time. Some of the earliest British Patents were for making pitch and tar from coal, and even 'essential oils, volatile alkali, mineral acids, salts and cinders' too. Yet in the youngest phase of the gas-making industry the by-product tar was considered to be 'an embarrassment'.

Dr Donald McNeil—a leading consultant on coal tar—describes the first established pattern of tar processing as a primitive distillation in cast iron pots to yield two products; they were a crude light oil used as a solvent in paints, varnishes and rubber solutions, and a refined tar. The latter was in demand as a protective paint, as a proofing agent for ships' cordage and for making lampblack. Later the separation processes were improved and refined so that from the middle of the nineteenth century the tar industry had a fast-growing demand for its chemical products. Leading chemists were carefully investigating the composition of the tar and its distillation products and preparing derivatives of interest notably in Britain and Germany with the leadership steadily passing to the latter towards the end of the century.

Like the coal industry, that dealing with coal tar and its derivatives has had many ups and downs. Its most recent highpoint was in 1957 when tar production reached a peak. In turn the industry for making gas from coal was hit first by the use of fractions from petroleum which were transformed into gas and then by the rapid development of the North Sea gas resources as more and more of Britain had its burners 'converted' to be suitable for the different characteristics of the methane-rich treasure tapped from under the sea.

Although 20 million tons of coal was still being carbonised in British gasworks as recently as 1964 (with a further 25·6 million tons going to coke-ovens), today virtually no coal is carbonised primarily for gas, and coal tar is derived mainly from coke-ovens operated to meet the need for coke in making iron. Consequently the amounts of coal treated fluctuate with the fortunes of that industry. In 1975/76 the steel industry, still suffering from its most severe recession in demand since the Second World War, produced less steel than back in 1955. Correspondingly, there was a setback in the amount of coal consumed in coke-ovens— down to 18·2 million tons. From these plants, coke is by far the most important product both in quantity and in value of revenue. It is, incidentally, largely used as a chemical, although not normally classified or considered in that way. For in the blast furnace—so-called since a blast of hot air is forced in under pressure—the coke burns, generating hot reducing gas which reacts with the iron ore chemically reducing it to iron. This makes coke the most heavily-used chemical in the UK or other iron-making countries. But chemicals in the more generally understood sense are also contributed on an appreciable scale from the carbonising of coal, despite the domination by petroleum sources.

When coal is carbonised

Generally, it is the carbonising of coal for making coke that is the main current source of coal-based chemicals. The coal is heated in tall narrow chambers made of silica brick, by gases at about 1300°C passing through ducts outside the chamber. Volatile products formed as the coal decomposes, pass off through an 'ascension pipe' and then through a purification and collecting train of pipes and towers. At the end of the period of treatment—between 15 and 30 hours—the red-hot coke is discharged into wagons. To economise in best quality coking coals, it is now normal to blend in coals of lower coking properties and finely crushed coke 'breeze' in accordance with the results of research. Batteries of coke ovens have in the past been the source of serious atmospheric pollution but increasing attention is now being given to cleaning up charging, operation and discharging to eliminate this nuisance.

Primary products are the gas, crude coal tar, ammoniacal liquor, benzole—largely a mixture of benzene, toluene, xylenes—hydrogen sulphide and some hydrogen cyanide, and of course the coke. By various processes the gas is stripped of the other volatile components which are then generally recovered and further treated so that they can be offered for sale. The apparently 'primary' products collected at the works are not wholly those directly resulting from the breakdown of coal under the influence of heating. For the latter are further decomposed after they have been formed and the conditions of carbonisation, the time spent by the true primary products in the oven and the method of recovery all influence secondary reactions and the physical condition of products.

The crude tars are known to contain many hundreds of individual chemical compounds. Some are recovered as separate chemicals but many of the commercial products emerging from the refining of coal tar are mixtures. The diagram, Fig. 1, reproduced from Dr McNeil's book, clearly shows both the primary fractions from distilling coal tar and the further separated constituents obtained from them. The final column gives a selection of the major industrial uses. Proportions given are only indicative since they vary with many factors including the coal and conditions at each stage of treatment.

The ways in which the chemicals of these fractions are converted into final products have been discussed earlier in Chapter 2. But even the murky residue from distillation—pitch—has a range of utility based largely on its unique combination of qualities. It is relatively cheap, adhesive, very waterproof and resistant to chemical attack. So, in combination with other materials, it forms the basis for building products for waterproofing, stopping leaky cracks and protecting metal structures against attacks by such corrosive agents as acids, alkalies, solvents, bleach, oils and fats.

I quoted earlier that coal is now reckoned to provide overall some 5 per cent of our organic chemicals. Within this overall figure, the proportions vary for individual chemicals. In 1975, for example, sales of benzene by UK manufacturers were given as over 83,000 tonnes from coal tar distillation and benzole refining compared with about 421,000 from other sources (in practice petroleum); for toluene, the same comparison showed about 19,000 with coal as source and 75,000 as the sales from other feedstocks.

Coal processing

In 1914 Bergius in Germany discovered how to convert coal into oil by the action of hydrogen under pressure and the German company IG built

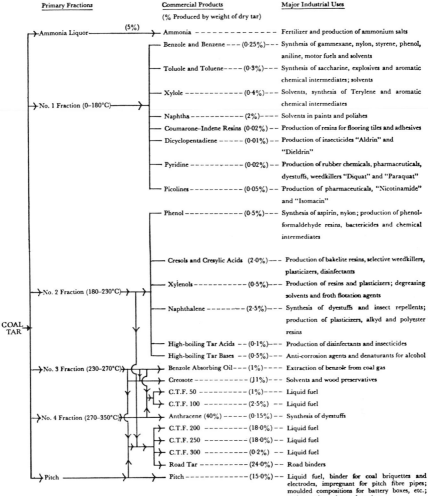

COAL TAR, ITS PRODUCTS AND THEIR MAIN USES

Primary Fractions	Commercial Products (% Produced by weight of dry tar)	Major Industrial Uses
Ammonia Liquor (5%)	Ammonia	Fertilizer and production of ammonium salts
	Benzole and Benzene (0·25%)	Synthesis of gammexane, nylon, styrene, phenol, aniline, motor fuels and solvents
	Toluole and Toluene (0·3%)	Synthesis of saccharine, explosives and aromatic chemical intermediates; solvents
	Xylole (0·4%)	Solvents, synthesis of Terylene and aromatic chemical intermediates
No. 1 Fraction (0–180°C)	Naphtha (2%)	Solvents in paints and polishes
	Coumarone–Indene Resins (0·02%)	Production of resins for flooring tiles and adhesives
	Dicyclopentadiene (0·01%)	Production of insecticides "Aldrin" and "Dieldrin"
	Pyridine (0·02%)	Production of rubber chemicals, pharmaceuticals, dyestuffs, weedkillers "Diquat" and "Paraquat"
	Picolines (0·05%)	Production of pharmaceuticals, "Nicotinamide" and "Isomacin"
	Phenol (0·5%)	Synthesis of aspirin, nylon; production of phenol-formaldehyde resins, bactericides and chemical intermediates
	Cresols and Cresylic Acids (2·0%)	Production of bakelite resins, selective weedkillers, plasticizers, disinfectants
No. 2 Fraction (180–230°C)	Xylenols (0·5%)	Production of resins and plasticizers; degreasing solvents and froth flotation agents
	Naphthalene (2·5%)	Synthesis of dyestuffs and insect repellents; production of plasticizers, alkyd and polyester resins
	High-boiling Tar Acids (0·1%)	Production of disinfectants and insecticides
	High-boiling Tar Bases (0·5%)	Anti-corrosion agents and denaturants for alcohol
No. 3 Fraction (230–270°C)	Benzole Absorbing Oil (1%)	Extraction of benzole from coal gas
	Creosote (J1%)	Solvents and wood preservatives
	C.T.F. 50 (1%)	Liquid fuel
	C.T.F. 100 (2·5%)	Liquid fuel
No. 4 Fraction (270–350°C)	Anthracene (40%) (0·15%)	Synthesis of dyestuffs
	C.T.F. 200 (18·0%)	Liquid fuel
	C.T.F. 250 (18·0%)	Liquid fuel
	C.T.F. 300 (0·2%)	Liquid fuel
	Road Tar (24·0%)	Road binders
Pitch	Pitch (15·0%)	Liquid fuel, binder for coal briquettes and electrodes, impregnant for pitch fibre pipes; moulded compositions for battery boxes, etc.; anti-corrosion paints and coatings

COAL TAR

Figure 1. Coal Tar, its products and their main uses. (From 'Coal Carbonization Products' by Donald McNeil)

a plant in 1927 to produce 100,000 tons a year of petrol from brown coal using a development of this process. Later, spurred on by war needs, output was steadily increased and technology improved until by the end of the war, hydrogenation plants were providing a large part of Germany's aviation fuel. In 1944 they produced a total of about $3\frac{1}{2}$ million tons of liquid fuels from coals, tars and residual oils from petroleum.

In Britain it was ICI which built a hydrogenation plant, in 1935—the first commercial hydrogenator for *bituminous* coal. In 1939 they stopped hydrogenating coal and switched to using a creosote oil as feedstock. From this they manufactured aviation petrol on an appreciable scale during the war and continued operating until 1958. As ICI explained in a memorandum to the Wilson Committee on Coal Derivatives in 1960, marked changes in costs and in market prices transformed the character of the financial results. When they decided in 1933 to build the plant there was a prospect of a petroleum shortage and a desire to develop new outlets for coal. Even then the returns expected were so poor that they needed the help of a tax preference on petroleum from indigenous materials to enable them to pay their way with this process. Estimates for conditions in 1960 showed that the operation would result in a considerable deficit. Both in the UK and in Europe the processes were abandoned (except, as the Wilson Committee noted, for a few small plants making specialised products).

For many years coke was an important source of chemical products. Between the First and Second World Wars the German production of synthetic ammonia, based on coke, at one stage met half the world's consumption of nitrogen compounds. In the Haber-Bosch process operated in Germany (and in Britain) the coke was reacted with air and steam to make mixtures of hydrogen, carbon monoxide (CO) and nitrogen. The CO was reacted with further steam in the presence of a catalyst forming carbon dioxide (CO_2) that was washed out with water under pressure. Nitrogen and hydrogen went forward to the ammonia convertors. Products emerging from this and associated plants yielded aqueous ammonia, ammonium sulphate, nitric acid, ammonium nitrate, urea and other compounds of great industrial importance. Methanol, too, was made from mixtures of hydrogen and CO.

Coke was also converted to calcium carbide. This was then decomposed with water forming acetylene—a highly reactive chemical building block; from acetylene, in the post-war years, industry produced vinyl chloride (for PVC), trichloroethylene, and acrylonitrile for acrylic fibres. In 1958 in the UK alone it was estimated that about $1\frac{1}{3}$ m. tons of coal were used for synthesis via the use of coke. From it were made about half a million tons of synthetic ammonia and almost 50,000 tons of acetylene. At about the same period, the USA was consuming about 400,000 tons a year of coke—a sixth going for metalworking and the rest as a chemical intermediate on the way to making solvents and the precursors for rubbers and plastics.

Broadly these processes were overtaken by cheaper chemical routes based on natural gas and petroleum sources particularly in the 1960s—although acetylene is still used for making vinyl chloride in the East European countries and, as discussed below, in South Africa.

Coal into gas

Coal can be converted almost wholly into gas leaving behind as solid (or molten) residue, the ash, and any desired proportion of the carbon. I intend to discuss this further in a later chapter since this conversion, known as 'gasification', may be one of the main ways in which coal is processed in the future. But a mixture largely of CO and nitrogen, called 'producer gas', made from coal or coke was an important fuel gas for many decades using a variety of plants ('producers'). Other commercial plants of several types have been working more recently based on reacting coal with steam forming hydrogen and the oxides of carbon. The reaction with steam absorbs heat and so heat must be generated or added to sustain the reaction. This heat is usually generated by adding air or oxygen to burn part of the coal—though techniques vary considerably.

The Winkler process—named after its inventor—treats crushed coal (in a so-called 'fluid bed' where the coal is suspended in a stream of gas) with oxygen and steam; originally it was devised for partial gasifying leaving behind active carbon but it is now operated to go all the way leaving only ash. It works best with reactive low-rank fuels like lignite (or coke from brown coal), but can also deal with other coals. The first commercial plant started up in 1926 and sixteen are now reported to be working in such diverse countries as Japan, Yugoslavia, Turkey and India. Some are making gas for synthesising chemicals and others making fuel gas. Existing plants work at normal atmospheric pressure, while developments are reported to be in progress to adapt the process to higher pressure.

The Lurgi process also gasifies coal by treatment with oxygen and steam but under a pressure of 20 to 30 atmospheres in what is called a 'fixed bed'—in practice a slowly descending mass of fuel undergoing conversion. This too started life converting lignite (some forty years ago) but was developed to handle bituminous coals and even anthracite. Fourteen industrial plants are known around the world, the largest being the greatly expanded installation in South Africa yielding gas for reaction. Elsewhere the gas produced—sometimes after enrichment to higher calorific value with oil products—has also been used as fuel.

Third of the currently-used types of plant, also developed in Germany, is called the Koppers-Totzek and is markedly different. At opposite sides of the reaction chamber are burners—one each side—facing each other. Fine coal is entrained in a stream of oxygen and steam (at roughly atmospheric pressure) and reacts at high speed in the chamber at higher temperatures than in the two previous cases, namely around 1800°C. Caking properties in coals that might otherwise cause complications in plant at lower temperatures, are here destroyed, tar and other volatile products are decomposed so that it is mainly CO and hydrogen that emerge, while about half the ash is fused to a slag. This is more or less an omnivorous piece of equipment consuming a wide variety of combustibles in solid, liquid or gas form. First commercial operation was in 1952 in Finland and the world tally of these plants is about the same number as for each of the two previously named.

These are now as a group referred to as first-generation gasifiers. Each has strengths and weaknesses; gases from them may be used directly as fuel, or enriched with other constituents as fuel, or as the basis for synthesis or for upgrading to a gas rich in methane as a fuel of much higher calorific value.

The largest chemicals-from-coal plant

Of all the approaches to synthesising chemicals from coal, probably most widely known by name and repute is the Fischer–Tropsch process, named after its two discoverers. In 1925 tnese two scientists found that a mixture of carbon monoxide and hydrogen could be made to react forming a mixture of hydrocarbons. Since then there have been major developments in catalysts, reaction conditions and processing technique. The two reacting gases may be produced by gasifying coal, coal products, or oil fractions or by breaking down natural gas with steam.

In Germany the first full-scale plant was built in 1936 by Ruhrchemie A.G. Together with further ones built in the following years, they eventually yielded about 8 per cent of Germany's home production of oil during the war and a smaller quantity of waxes, lubricating oils, detergents, synthetic soap and fat. There were further process developments in Germany after the war.

At various times, projects or plants have been reported in Japan, France, Poland, USSR, India, China and Australia. The USA has shown fluctuating degrees of interest and has projects in its extensive current programmes of energy R & D. Britain has never had a commercial Fischer–Tropsch plant, but a 'main item' listed in the programme of the Fuel Research Board when it was set up in 1917 was to investigate methods of producing liquid fuels from coal. At the former Fuel Research Station in Greenwich, experimental work on the Fischer–Tropsch process started in 1936 continuing later at the Warren Spring Laboratory in Stevenage; but the programme closed down in 1961.

The only major commercial Fischer–Tropsch plants in the world are those in South Africa using coal that is exceedingly cheap—its low price being attributed to favourable geological conditions but presumably also connected with particularly low wages for its miners. The first of these was completed in 1955 and the South African Coal, Oil and Gas Co. Ltd.—briefly known from the initials of its Afrikaans name as SASOL—has since expanded its chemical-from-coal activities. In both the first plant (Sasol I) and in Sasol II, the coal is gasified in Lurgi equipment by reaction with steam and oxygen under pressure. In Sasol I the gases are then divided and part passed through a fixed catalyst reaction bed, the balance going to a fluidised, entrained bed reactor. The later, much larger, plant now being completed uses only the entrained bed system, known as the Synthol process. From the fixed bed reactor, about one third of the liquid products can be used as petrol; the rest comprises straight chain high boiling hydrocarbons, medium oils, diesel oil, a range of waxes, some alcohols and ketones. Liquid products from the entrained bed process contain over 70 per cent of petrol—broadly products with carbon chain lengths in the range of C_5 to C_{11}—with higher proportions of low boiling hydrocarbons, more alcohols and ketones and more aromatics. There is much less waxy, high-boiling material.

Preliminary estimates of the output from the newer plant include such items as $1\frac{1}{2}$ million tons a year of motor fuels, 160,000 tons ethylene, 100,000 tons ammonia, almost as much sulphur and many other products.

In a review of 'the coal renaissance', naturally with major attention to the plants briefly discussed above, Mr P.E. Rousseau, Chairman of SASOL, also drew attention to other chemicals-from-coal activities in his country. Sasol I makes 20,000 tons a year (on a nitrogen basis) of ammonium sulphate from gas liquors and a further 50,000 tons syntheti-cally from the gas rich in hydrogen emerging from the oil synthesis plant. Other manufacturers are making similar quantities by gasifying coal in the Koppers-Totzek gasifiers and using the gas for synthesis. While the US and Western Europe has generally abandoned the use of carbide as a source of acetylene for making vinyl chloride for PVC, a new project near Sasolburg will be based on this chemical route—the carbon coming from coal—and have a capacity of 100,000 tons a year of PVC.

Metals from coal?

Very often you can see in coal grey, white or golden streaks that are the signs of the mineral matter. Both these and the finely-divided invisible components arise from

* The small amount of mineral matter in the original plant material;
* silt blown or washed into the decaying organic matter;

BY-PRODUCTS PLANT

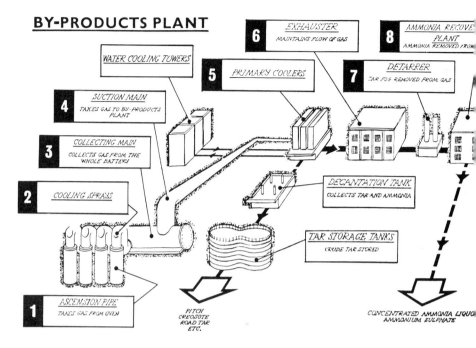

Figure 2. By-products plant. (NCB).

* dissolved salts in the waters of the swamps or those later pouring over the decaying organic matter;
* waters trickling through the coal as it was being formed.

For most uses of coal, these parts are purely a bane.

In industrial combustion plant, they force a need for equipment to trap the residue. Both in industry and the home they dilute the coal, producing ash that has to be removed; they may trap carbon during burning, aggravate the fouling of equipment (notably the cooler parts of boilers) and/or cause erosion, particularly when coal is burnt and used to drive gas turbines. On the credit side can be mentioned that fly ash—the fine particles entrained in flue gases when powdered coal is burnt—is being used for cheap road-fill and for making bricks, concrete and lightweight aggregate.

But in terms of our main theme, the incombustible part is of potential interest because of the very wide range of metals already mentioned. Most are in trace amounts but aluminium and iron are found in relatively larger concentrations. Processes have been suggested for extracting these metals from the shales associated with coal. A more ambitious project has been a

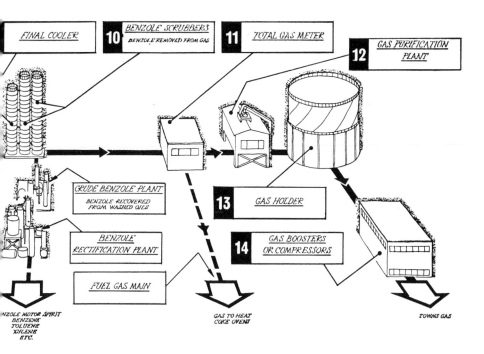

design for an integrated operation yielding aluminium, iron and potash from colliery waste, but assessments have shown this to be uneconomic.

Germanium—an element used in electronic devices and as a catalyst and in special glasses—has been recovered and at various times there has been research, sometimes leading to working processes, for selenium, gallium, mercury and even the conventional precious metals gold and silver. But interest in these areas waxes and wanes according to current trends in demand, supply and market prices. Even when these projects have gone ahead they have been of very minor commercial significance in terms of coal chemicals.

Current Developments

What is being done to increase the yields of chemicals and other conversion products from coal? For Europe in the short-term, the neatest comment was that by Dr David Dainton, Director of the NCB's Coal Research Establishment (CRE). 'In the short-term in Europe' he wrote 'the sensible thing to do with coal is to burn it. The cheapest way to make

E

oil from coal is to increase the proportion of coal used in combustion and release natural oil for more esoteric purposes'. Behind this comment is the fact that many of the coal conversion processes under development with a longer-term perspective yield products that are currently dearer than those already on the market.

Nevertheless there are shorter-term possibilities too.

Much of the research at CRE itself is devoted to making or improving coal derivatives—generally from existing carbonisation processes—for early marketing. Products have included fire retardants based on ammonium sulphate for treating timber and paper, and variants suitable for using in fire-extinguishers. From the phenols of tar, they have developed a closed-cell foam board of phenol-formaldehyde to meet the demand for thermally insulating roofs. Phenolic resin has also been produced in a form suitable for a varnish with high electrical and mechanical qualities. Yet these are projects for upgrading or extending the range of use of existing products.

In addition, throughout the world the R & D programmes of the major coal producing countries have also tended to include projects aimed at increasing the yields of conversion products by modifying pyrolysis and associated processes. Broadly the aim is to escape from a key problem of conventional coal coking—that the demand for the coke controls the supply of chemicals from coal. Dominating all others in magnitude are those of the USA. There, coal processing developments are being pressed forward with the urgency that comes of knowing that the energy crisis has already begun and the main hope of reducing dependence on imports lies in making use of the enormous indigenous reserves of coal. The work is heavily supported by the government, through the Energy Research and Development Administration (ERDA).

Among their projects is the multi-stage pyrolysis COED process which was successfully operated on a scale of 36 tons a day by the FMC Corporation. This meets the basic requirement for improving the yield of liquid products—carbonising for short periods (minutes or in other processes, seconds) with very high rates of transferring heat to the coal. It uses fluid-bed techniques—discussed further in Chapter 6—and has been tested with a wide range of coals. Though the original plant was closed down in 1974, designs have been prepared for developments of the process using up to 25,000 tons of coal a day, with a target date of 1980. Studies of the synthetic crude oil produced have shown that it has the same value to a petroleum refiner as a typical US 'sweet' crude oil with only 0·3 per cent of sulphur. The conclusion resulted from an evaluation of process yields and product quality in an extensive pilot plant programme.

Among the wealth of other processes at various stages of development, is one known as Coalcon which hydrogenates as well as pyrolysing coal under pressure yielding 51 per cent of liquids and 49 per cent of substitute natural gas (SNG). Due to convert 2,600 tons of coal per day, this plant was scheduled to operate in February 1979, but ran into engineering difficulties. In Germany the Lurgi-Ruhrgas rapid carbonisation process is being applied in plants converting 1,600 tons a day of lignite. Other countries also have programmes for fast carbonising at an advanced stage. For example the Australian Commonwealth Scientific and Industrial Research Organisation have a flash pyrolysis plant working on a tonnage-scale aimed at producing a crude oil.

In Japan, the Osaka Gas Co. has developed a process with a different emphasis called the Cherry-T, an acronym derived from Comprehensive Heavy Ends Reforming Refinery. Here the aim is to improve the efficiency of treating coal tar in producing pitch. The tar is retained in a reactor to promote condensation and is discharged at about 400°C, before distilling off the oil. The three major uses of the pitch that remains are as coke moulding material and as pitch for electrodes and for pitch coke: the main user of the latter two products is the aluminium industry. Pilot plant variants of this have been reported where coal powder is added to crude oil before this reaction stage.

Naturally there are many other programmes at various stages of scale and development throughout the world, including an expanding one at CRE, intended to convert some or all of coal into more valuable commodities and having a medium or longer-term perspective. This means that there is necessarily an overlap between the work discussed above and the subject matter of Chapter 7—but the latter looks more to the developing future.

Further Reading

'Coal Carbonization Products'. Donald McNeil. Pergamon Press, London.

'Carbonization of Coal'. J. Gibson & D. H. Gregory. Mills & Boon Ltd, London, 1971.

'Blast Furnace Coke—Macro and Micro Considerations'. L. Grainger. The Coke Oven Managers Association, 1974.

Report of the Committee of Coal Derivatives. 1960, Cmnd 1120, HMSO (known as the Wilson Committee Report).

'British Steel Corporation (Chemicals) Ltd' (British Steel Corporation).

'Smokeless Fuels and Tar' (National Coal Board).

'The Robens Coal Science Lecture, 1975: the coal renaissance; a South African point of view'. P. E. Rousseau. J. Inst. Fuel, 167, pp 167–175, Dec. 1975.

Business Monitor, PQ 271, Quarterly Statistics. General Chemicals, HMSO.

Quarterly Bulletin of Energy Statistics. Statistical Office of the European Communities, EEC.

'Trace Elements in Fuel'. A symposium sponsored by the Division of Fuel Chemistry at the 166th Meeting of the American Chemical Society, Chicago, Ill, 1973. American Chemical Society, 1975

'Co-production of Iron and Aluminium—A New Pattern for the Steel Industry'. D. L. Levi. Symposium on mineral wealth. British Association for the Advancement of Science, 1973.

'Coal Processing'. R. J. Morley. Article in Encyclopaedia Britannica, 1974.

'Fossil Energy Program Reports' of the US Energy Research and Development Administration.

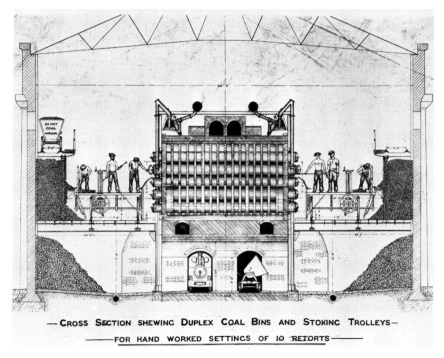

— CROSS SECTION SHEWING DUPLEX COAL BINS AND STOKING TROLLEYS—
————FOR HAND WORKED SETTINGS OF 10 RETORTS————

Detail from an early print showing hand worked retorts at Vauxhall gasworks. (Photo. Science Museum London).

Chapter 5

Energy from coal—past and present

A great 'smog'—a smoke-laden fog—blanketing London from the 5th to 9th December 1952 inclusive was estimated to have caused the deaths of some 4,000 people, that is, deaths in excess of the numbers normally expected to die over that period. Deaths from bronchitis increased by about nine-fold during the week ending on 13th December and from pneumonia about four times. There was enormous pressure on hospitals due to big increases in the numbers suffering severely from respiratory diseases and heart disease.

The diagram (Fig. 1) shows the degree of correlation between the extent of pollution by smoke and sulphur dioxide, and the daily death rate in Greater London. There was an abnormally high death rate in the area for several months after this disaster (Fig. 2). But in this case at least it can truly be said that the victims did not die in vain. For the results so shocked public and government that an expert Committee (the 'Beaver' Committee) was set up to examine causes and effects of air pollution, and to recommend preventive measures. And, unlike many other such bodies before and since, this Committee did indeed make specific proposals resulting in a whole sequence of events including Clean Air Acts and later Acts that tightened up considerably the laws protecting the environment.

Fogs, pointed out the Committee, are natural phenomena not subject to human control. What made them dangerous was the combination of natural fog with pollution. And the smoke in the air aggravated the risks to people by making these dangerous fogs more frequent and more persistent, providing nuclei for moist fog particles and cutting off sunshine that would otherwise disperse them. The smoke and grit were considered to come wholly from bad methods of burning coal; most of the sulphur dioxide and carbon monoxide also came from coal. It should be added immediately that in the intervening quarter century the patterns

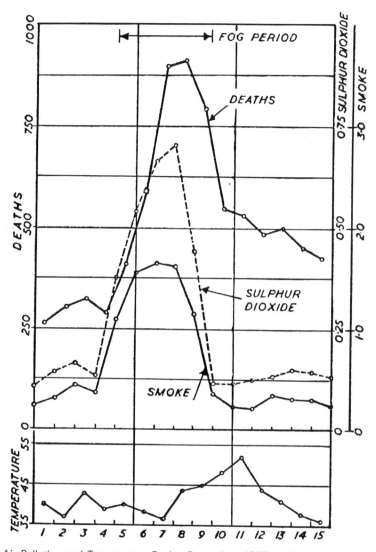

Deaths, Air Pollution and Temperature During December, 1952, in Greater London (Population 8½ millions).

Units—Sulphur Dioxide...   Concentration in parts per million parts of air* (mean of ten sites).

      Smoke . . . .   Concentration in milligrams per cubic metre of air (mean of twelve sites).

      Temperature . . . .   Degrees Fahrenheit (noon readings on Air Ministry roof).

      Deaths . . . .   Total number occurring each day.

*Note: 1 part per million parts by volume of air equals 2·86 milligrams of Sulphur Dioxide per cubic metre of air at N.T.P.

Figure 1. From 'Interim Report of the Committee on Air Pollution' Cmd 9011. HMSO 1953.

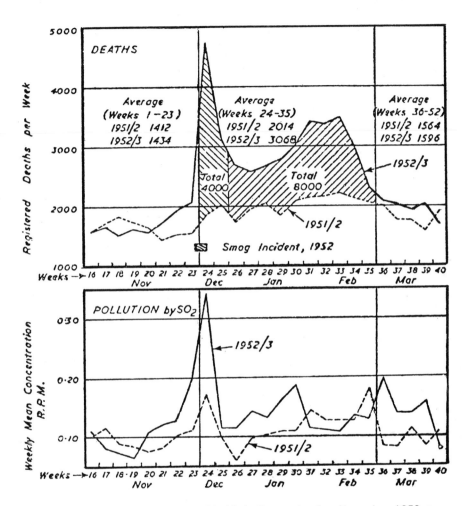

Deaths and Air pollution by Sulphur Dioxide in Greater London, November, 1952, to March, 1953.

Units—Sulphur Dioxide... Concentration in parts per million parts of air.
Deaths .... Number registered each week.

The weeks are numbered consecutively 1–52 from July, 1952, to June, 1953. The period shown in the diagram (weeks 16–40) covers the period from mid-October to the beginning of April. The smog incident of 5th to 9th December, 1952, overlaps weeks Nos. 23 and 24.

Figure 2. From 'Interim Report of the Committee on Air Pollution' Cmd 9011. HMSO 1953.

Figure 3. Widnes towards the end of the nineteenth century. (Courtesy of ICI).

have changed in many respects that are discussed further below. One very important difference is that chronic bronchitis is now found to be closely related to numbers of cigarettes smoked and not to fluctuations in the much lower levels of air pollution of today.

However a large part of our heritage of blackened houses and other buildings was undoubtedly due to the bad styles of burning coal in the past (Fig. 3) Because it is so easy to make a fire, people have under-rated the need for a research and design effort to eliminate smoke and grit, reduce invisible gaseous pollution and give the user the maximum benefit from the energy stored chemically in the coal. This energy ultimately derives of course from radiant energy coming from the sun promoting the growth of the original plant products by the process known as photo-synthesis. In recent years there has been a marked growth in understanding that we should treat these precious stores of accumulated solar energy with respect since they are formed so slowly and—large though they are—remain ultimately limited. But even before the soaring price of oil, particularly in 1973/74, administered a sharp shock to a world that had assumed fossil fuels to be cheap and abundant, public pressure, a succession of laws and campaigning by the National Society for Clean Air (or its

Sheffield before the Clean Air Act.

Sheffield City Centre today.

forerunners) had resulted in increasing control over offenders (including the chemical industries in their earlier stages) and a steady improvement in the cleanliness of the air we breathe.

Illustrations given in a booklet by the Department of the Environment called 'Clean Air Today' refer to several different kinds of evidence.

'Visibility between December and February in central London now [1974] averages $4\frac{1}{2}$ miles. It was only $1\frac{1}{2}$ miles in the late 1950s. Midwinter sunshine in central London is now 50 per cent greater than it was ten years ago.

There are today many species of birds, eg swifts, and more types of butterflies in the London parks than there were only ten years ago. Trees grow more healthily, birds and flowers which a generation ago had apparently become extinct in most parts of central London are now returning to the City's parks.

'People, too, have fewer chest colds . . . we are now able to clean our historic buildings.'

The relevance of these profound changes to our main theme is that the principal offender was for many years the domestic chimney discharging at low level the products of inefficient burning of coal in the open grate. Locally, of course, in industrial areas, there have been major nuisances caused by other sources. But overall it has been the consequences of decomposing coal on home fires rather than thoroughly burning it, that has caused most of our pollution troubles for many years—in terms of smoke, grit and dust.

For the year of the catastrophe, 1952, the Beaver Committee estimated that over two million tons of smoke had been discharged into the air. About a fifth of all the coal was burnt on domestic grates and yielded almost half of the smoke. The incomplete combustion also invisibly produced the poisonous gas carbon monoxide in large quantities. There were contributions to this fouling of the atmosphere from oil sources too but they were relatively small. At that time, the latter represented a relatively small part of the national fuel usage. Its main defects in this context were the high proportion of sulphur dioxide it yielded (though still a small quantity) and the dense filth emitted from badly tuned diesel engines.

What were the recommendations for preventing a repeat of the major lethal event and more generally, for cleaning up Britain's air? They were divided into those entailing legislation and others. The former comprised such features as

* prohibiting the emission of dark smoke;
* making it compulsory to have efficient grit and dust arresting plant on new industrial installations;

* giving power to local authorities to establish smokeless zones and smoke control areas where the use of bituminous coal for domestic purposes would be restricted;
* providing financial assistance towards the costs of converting grates in smokeless zones and smoke control areas;
* demanding annual reports on smoke abatement from local authorities to the relevant minister.

The other recommendations covered some of the logical consequences. The British Standards Institution was to set up codes of good practice and draft relevant specifications on a number of technical issues, and industrial stokers were to be properly trained and paid. Supplies of smokeless fuels should be increased, and also the equipment to burn them in. Clean air should become a 'declared national policy', development and research started or accelerated and a Clean Air Council set up to push matters along in this new area of national concern.

And so it came to pass. For the first Clean Air Act became law in 1956 and has since been supplemented by other laws, and there is a Clean Air Council. The result has been dramatic. Some aspects have already been quoted above. Emission of smoke was estimated to be down to 0·39 million tons for 1975—about a fifth of that in the year of the Great Smog—still dominated by the filth passing up home chimneys but enormously reduced in quantity. Industry, meanwhile, was being given an almost clean bill of health.

How has this revolution been achieved?

The formal steps towards controlling air pollution have been listed by the National Society for Clean Air. Among them are such events as these:

1954 Final Report of the Beaver Committee
1956 Clean Air Act
1959 International Diamond Jubilee Clean Air Conference
1966 First Congress of the International Union of Air Pollution Prevention Associations
1968 Clean Air Act
1971 Royal Commission on Environmental Pollution
1972 UN Conference on the Human Environment
1974 Control of Pollution Act
1974 Health & Safety at Work Act

Cleaning up the atmosphere

This indicates the pressures. But how have industry and the domestic user actually reduced the discharge of smoke into the air? Immediately before the Second World War it was estimated that domestic grates were pouring out about $1\frac{3}{4}$ million tons of smoke a year and other sources a further million. Above we quoted the assessment by the Beaver Committee for 1952 (1 million tons domestic, 1 million others), and then the figure for 1975 (which comes from a report by the Warren Spring Laboratory). The 0·39 million tonnes total was composed of 0.35 million from domestic heating and only the incredibly low figure of 0·04 from industry including railways.

Over this period there were marked changes in the pattern of fuels consumed, but also a great improvement in the efficiency of using coal in industry particularly after the Clean Air Act came into effect. Over much the same period coal lost ground as a fuel, while oil and natural gas grew rapidly in importance.

Dr A. Parker, former Director of the Fuel Research Station which used to be at Greenwich, has summarised the reasons for decreased smoke from households under these headings:

1. During the war period coal and coke were rationed. Householders began to use more electricity and gas for heating and decided that these were more convenient and cleaner than heating by coal. This, he implies, set a trend towards more convenient fuel.

2. The Clean Air Act enabled local authorities to establish smoke control areas in which the burning of bituminous coal in open domestic grates is prohibited.

3. Since 1945 there has been an increasing number of housewives taking employment; this has further increased the demand for convenient gas and electric methods of heating.

Certain it is that sales of bituminous house coal dropped sharply from $30\frac{1}{2}$ million tons in 1956 to only 7 million in 1975–76. Yet the rise in cost of oil, combined with a drive to sell smokeless domestic fuels and new efficient appliances for the solid fuels is thought to be ending the decline in using solid fuels in the home without any detriment to the quality of the environment. During 1977–78 there were increases in sales of domestic coal and of equipment approved by the Domestic Solid Fuels Appliances Approval Scheme for the NCB. These include open fires, inset or freestanding, sometimes with an underfloor air supply, some with boilers, as well as the stoves known as roomheaters, and improved cookers and boilers which often have thermostats and may have timer controls. Generally they are designed to burn a variety of manufactured smokeless fuels—mainly from low temperature carbonisation—or natural ones, such as stated grades of anthracite or the coals of low volatile matter called Welsh dry steam coal.

With contemporary good looks to form an attractive focal point this Redfyre Plus 6 solid fuel room heater provides central heating for a 3-bedroom house with a large living/dining room. The total output of 36,000 Btu/hr. is balanced to give 10,000 Btu/hr. radiation and convection to heat rooms up to 1500 cu. ft. while the backboiler (26,000 Btu/hr.) heats 6 radiators and hot water for the house. With either manual or thermostatic control the burning rate is regulated by the heat-resisting dial set beneath the door panel so that all but the setting in use is concealed.

Great efforts have been put in to making appliances both easier to operate and attractive in appearance. Improvements include concealed rotary ash boxes needing emptying only once or twice a week, and effective controls over rate of burning. The latter mean that a fire can be left 'slumbering' for a period and can then be restored to full heat output in a few minutes notably when a fan forms part of the design.

Possibly the most revolutionary in concept are those known as 'Smoke Eaters'. In these the air flow is made to pass downwards through the fire-bed (Fig. 4). Using cheaper grades of coal which would normally burn generating smoke, the Smoke Eaters are designed to complete the combustion of the smoke in a secondary chamber where it is mixed with heated secondary air. Consequently only clear flue gas passes up the flue. The closed roomheaters applying these principles tend to be used for full central heating systems; they incorporate boilers and can run several

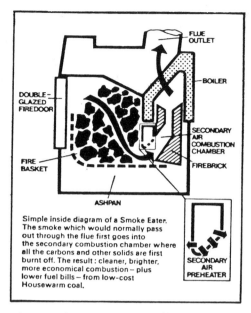

Simple inside diagram of a Smoke Eater. The smoke which would normally pass out through the flue first goes into the secondary combustion chamber where all the carbons and other solids are first burnt off. The result: cleaner, brighter, more economical combustion – plus lower fuel bills – from low-cost Housewarm coal.

Figure 4. How a modern roomheater gave up smoking.

radiators as well as providing hot water for the house. Because of the unique principle, they can be installed in Smoke Control areas although they do not need to burn smokeless fuels.

The NCB's Coal Research Establishment (CRE) carries out tests of these appliances not only for the NCB itself, and also for the Department of Industry and for private manufacturers, under the Approval Scheme quoted above. Fuels too are tested. Principal aims are to verify that fuels burn satisfactorily without making smoke, and in the case of appliances to determine the efficiency. For a smokeless fuel roomheater with back boiler a typical value is 75% for overall efficiency while a modern open fire with wrap-round back boiler would achieve 70%.

Another principal former source of smoke was the railways. When trains were hauled by steam-driven locomotives, the inefficient combustion on the grates under the engine boilers sent large volumes of smoke into the air. Back in 1938 13·6 million tons of coal was used by the railways at average efficiencies estimated to be only a few per cent. These locos have been replaced by diesel and electric engines; the former simply eliminate the use of coal (though sometimes themselves making smoke), while the latter ensure that the coal is used at higher efficiencies and burnt without producing smoke.

The Parkray Coalmaster is the second roomheater in the smoke eater family, and will heat up to 8 radiators plus providing ample hot water. Like all smoke eaters the running costs are extremely low.

Using waste heat

The country's largest user of coal, the Central Electricity Generating Board, taking 75 million tons in the year 1976/77, reports increases almost every year in the efficiency of converting the primary energy of the coal. From a level of around 28 per cent a decade earlier, average thermal efficiency rose to over 31 per cent in 1976. It is now widely understood that this low figure is not due to technical defects within the process of generating electricity. Steam is formed in the boiler at very high efficiency but in the following stage only about half of its energy is converted into work in the turbine—for fundamental reasons of basic physics. The final stage, converting work into electricity is again highly efficient. The visible evidence of the thermal inefficiency of the middle stage is the discharged heat that has to be deliberately wasted or dissipated in cooling towers or in warming our rivers. Understandably there has been increasing attention to the possibilities of recovering this heat. For example the CEGB have been experimenting in heating greenhouses with power station cooling water—collaborating with the Ministry of Agriculture, Fisheries and Food, and similar work has been reported from Germany in cultivating field crops. CEGB are also collaborating in experiments on using this low-grade heat in fish farming.

But one of the great dazzling visions in this area has always been the prospect of combined heat and power (CHP) applying the waste heat for use in district schemes by homes, shops, schools, hospitals and offices. To examine this, the Department of Energy through its Advisory Council set up a special Study Group. After considering the economics they reported that prospects for such domestic and commercial applications were not good in the short to medium term.

Power stations, they noted, discharge their heat as water that is only lukewarm. By changing the method of operating stations, they could be made to yield hot water say at 80 to 120°C as needed for district heating. This would reduce the electricity production but could increase the efficiency of using the fuel from about a third to 80 per cent. So CHP can certainly save energy. But the issue examined was whether this saving could be made economically taking into account all the costs of producing, transmitting and distributing the heat. Their conclusions were reached after visiting successful CHP schemes in Sweden, Denmark, Germany and France, and noting the differences in circumstances between the countries. One of the important ones was the housing density.

Small schemes, they concluded, were definitely not economic. Large CHP stations serving big cities showed better promise but still provided 'no immediate economic incentive' unless 'fuel costs rose substantially in real terms or if a lower discount rate was used for the assessment'. The one area that did offer 'good economics' was the development of nuclear-powered CHP schemes. In the present style of the Department of Energy, the report was published with an invitation for people to comment and criticise; it also referred to associated social, organisational and technological problems that needed thought alongside the essentially economic study. Among them were the need for 'high market penetration' for CHP schemes. How could this be reconciled with consumer choice, and with existing public utility services already supplying fuel? So, for the moment, at least the domestic and commercial side is regarded as a rather doubtful prospect.

Yet in the longer term CHP schemes could prove attractive, particularly if fuel prices rise in real terms. They would need a long development period and early decisions are needed if CHP is to play an important role by the turn of the century. This in turn demands an assessment of the density of the heating load around the country. So studies have been set up in five areas of Britain to survey building types and floor space so that the pattern of this heating load density can be determined; the patterns will be built up for each area, then extrapolated for all urban areas in the UK.

On the other side of this important debate, spokesmen of the electricity generating industry add points about the difficulties of 'balancing' demand for electricity and heat in schemes of these types. For it is evident that efficiency is highest when the heat and the electricity are used at rates that correspond with each other. In practice both vary from hour to hour, from day to day and particularly change markedly with the seasons. The discrepancy tends to be worst in summer when there is a smaller fall in demand for electricity than the large decline in demand for heat. And a cold weekend in winter can mean a large demand for heat though there is a relatively low call for electricity when most of the factories are not working.

A further complication arises from the way that the electricity industry uses the most efficient stations for the maximum time possible; the less efficient ones are run only as needed to meet higher demand. So, generally, as a station becomes older, it tends to be run for ever shorter periods. How can this pattern of use based on criteria of maximising efficiency in producing electricity be reconciled with the continuing need to serve say a district heating scheme? It is these kinds of considerations that leads them to be sceptical about the practical likelihood of achieving the fuel savings of some 45 per cent that at first glance appear to be available by linking together provision of electricity and of heat.

Understandably it was estimated by the Department of Energy that in 1977 less than one per cent of the space heating load of the UK was supplied by district heating schemes and very little of that from CHP plant. Probably the largest network is that at Nottingham supplying the city centre and 7,000 houses though most of the group heating schemes cover fewer than 200 houses.

But in industry—where transmission distances are much shorter and the power generation, power use and heat use may all be under the control of one management—there is a growing tendency to use heat recovery systems with electrical generators. Early in 1978, the first CHP station to be built by the electricity supply industry in the UK was approved. An estimated 13 MW of heat will be supplied to industry from engines driving two generators totalling 15 MW electrical. Overall efficiency is estimated at 76 per cent, by the Midlands Electricity Board. These power plants linked with means of using the heat for processing and space heating are sometimes called Total Energy (TE) systems. A few years ago they were reckoned to generate well over a fifth of the industrial power requirements of the UK. There are three main types of TE system.

1. Steam is generated at much higher pressure than is needed on the

plant. It drives a steam engine, usually a turbine which turns a generator making electricity, while the discharged steam goes on for use in the process. Systems of this kind have been long established, fired by coal or oil.

2. Fuel—normally selected liquid fuels or gases—can be burnt to drive gas turbines, again linked to generators, and the gases discharged which may be at say 500°C can be used for drying or as preheated air for other purposes.

3. When the principle is applied to a reciprocating internal combustion engine (for generating the electricity) discharged heat can be used from both the engine exhaust gases and the engine cooling water.

Such systems have been the subject of technical conferences (mentioned in the reading list at the end of the chapter).

In addition, a more recent variant of the second option is gasifying coal for power generation in combined cycles. According to a recent review the only large-scale operational plant of this type is at Lünen in the Federal Republic of Germany, based on a Lurgi gasifier. Another is due to be built at Pekin, Illinois, and there are many others under development. This approach demands gas turbines that can withstand inlet temperatures around 1200°C but also requires very clean gas to avoid fouling, corrosion or erosion—thus setting high demands for gases generated from coal.

Among the organisations working on this is CRE. They point out that the gasification process must be efficient, capable of handling a wide range of coals and provide gas that will not damage turbines. To meet these needs they have been running experiments on equilibria and rates of gasification reactions. It has proved necessary to keep a watch on the fuel gases produced, for alkali metal salts and sulphur compounds. But gas of acceptable calorific value has been manufactured that could be used in gas turbines generating electricity and operating in conjunction with steam turbines to give improved thermal efficiencies. In 1977, there were known to be three further teams—in the USA—working on processes using fluid beds to make gas of relatively low calorific value suitable for power generation.

Can fuel cells save us fuel?

A fuel cell is a highly efficient device for converting the chemical energy of fuel into electricity. It has an anode where fuel is oxidised, a cathode where an oxidising agent (usually oxygen) is chemically reduced and an electrolyte; this separates anode from cathode and conducts the required electrically charged particles (ions) between them. In applications where cost is no object and technical efficiency is paramount—as in the moon shots—they have proved very successful.

But for more workaday applications they have so far proved expensive and far too short-lived, due to poisoning of the electrodes. Early pioneering work in the UK led to the Bacon cells used in the Gemini and Apollo space programmes; now a number of research projects are in progress throughout the world intended to lead to economic and practical fuel cells. In the development programme of the US Institute of Gas Technology (IGT) the aim is to produce in ten years, fuel cells with up to 50 per cent efficiency in converting oil and coal to electricity; even higher efficiency is expected when the discharged heat from the reaction within the fuel cells is recovered in integrated systems. Overall, the objectives are savings of half of present fuel requirements per unit of electricity produced; the savings will come from improved efficiency in converting fuel energy into electricity, using waste heat and eliminating transmission losses that arise in distributing electricity from big central stations. Fuel cells—it is envisaged—will be used in smaller generating units providing both electrical and thermal energy for medium size users such as shopping centres, building complexes and groups of factories.

Cells of this kind will of course be usable with a range of fuels but since coal will steadily gain in importance, an important part of the development is that directed to applying fuel gases from coal. The fuel cell yields direct current which is then fed to an inverter to yield alternating current that can be supplied to the grid. The graphs (Fig. 5) show how relatively small fuel cell power plants have higher efficiency than even large steam and gas turbine systems, justifying the approach based on small to medium size units.

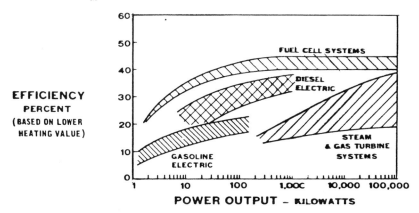

Figure 5. The full-load efficiency of fuel-cell power plants is high over a wide range of plant sizes. (Institute of Gas Technology).

What of MHD?

Systems based on magnetohydrodynamics—mercifully known briefly as MHD—produce electricity without using any moving solid parts in the generator itself. Hot fuel gas is ionised (electrically charged) and passed through a power channel surrounded by a magnetic field. Electrodes suitably placed in this gas stream can tap off electrical power.

Though the first model was built forty years ago relatively little success has been achieved so far with this approach. Yet it is considered to have the potential for generating electricity at efficiencies very much higher than current power stations and to lend itself readily to linking with a turbine generator to use the waste heat from the MHD unit.

The USA and the USSR have co-operated in joint programmes. Latest and largest of these will be based on a 25 Megawatt test unit to be located in Moscow. The gas may be produced by burning coal, natural gas or oil.

How efficiently do we use energy?

There is surprisingly little detailed information on how well we use our primary energy. In its Annual Digest of UK Energy Statistics, the Department of Energy approaches the issue on a national scale by providing tables based on the use of energy year by year and also on growth rates for the economy and for energy use. In the year-by-year table, ratios are derived for the amount of coal (or coal equivalent) consumed per £1,000 of Gross Domestic Product valued in constant terms at 1970 factor cost. This indicates a reasonably steady improvement in the energy used to produce our goods and services for the figure declines from 8·6 tons of coal equivalent needed to yield £1,000 pounds worth in 1955 to a corresponding 6·8 twenty years later. It suggests an improvement of over twenty per cent (Fig. 6). However the further table based on growth rates for both primary energy consumption and for Gross Domestic Product (again of course at constant cost) gives a more confused picture. Here the energy coefficient used is the ratio of these two growth rates and the figures fluctuate wildly making it difficult to discern any clear trend using this ratio as a criterion.

Even for sectors of the economy or for individual firms, it has still proved difficult to determine clearly the trends in efficiency of using fuel. John Chesshire and Christopher Buckley, research workers in an Energy Project team of Sussex University, interviewed large industrial energy users in late 1975 to investigate factors affecting patterns of using fuels. They naturally hoped to assemble, among other features, a fairly detailed picture of trends in consumption. They found a surprising 'paucity of

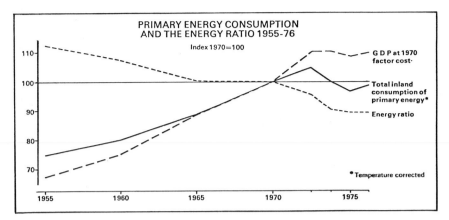

Figure 6. The energy ratio relates the level of total energy consumption to Gross Domestic Product (GDP) (From 'Economic Progress Report' July 1977. Prepared by the Treasury).

industry's energy data' although they had approached the largest companies in each sector, that is those where one would have expected to find that the necessary records were being maintained. This they explain in these ways

* Apart from the few energy-intensive sectors—iron and steel, cement, aluminium, glass and transport—energy typically accounts for under 5 per cent of total production costs. So energy questions received little attention.
* The companies had grown through acquisitions and changed their activities making it difficult to compare like with like in energy-output relationships.
* Product and process innovations had led to important structural shifts. In some cases the changes had meant energy saved for the company under investigation, although the raw material (or intermediates) represented extra energy used 'upstream' in the supply chain.
* Even where companies did keep records, they were noted in the original terms of measurement such as gallons of oil, tons of coal, therms of gas and kWh of electricity. Only a few had standardised the information and aggregated it on a consistent basis of energy units that would allow them to assess trends when fuels had been changed.

For these reasons the researchers discovered that companies could only rarely provide data in a form that made it possible to analyse trends in energy use related to output. The exceptions were those with the sharpest economic spur—those in the energy-intensive industries, who could quite specifically show 'substantial energy savings per unit of output'.

As a further exercise Chesshire and Buckley had wanted to make the same analysis for detailed sectors of industry. They were hindered by the fact that the sector analysis in the Digest of Energy Statistics had a breakdown for only 10 major industrial sectors; thus the data were more aggregated than in the official output statistics. On this limited sector analysis they were able to confirm that the energy-to-output ratio on a consistent basis had varied little in industries where energy consumption was not a major expense—such as engineering—but showed a big improvement in those with high energy expenditure. For the group bricks, tiles, china, glass and cement, for example, this ratio fell from 109·6 in 1964 to 78·3 in 1975. A surprising finding was that the group textiles, leather and clothing had also made a big improvement, though one would not think of these as major fuel users; but they were the only exception to the generalisation that heavy users made the greatest efforts.

Broadly, the others—where they did at all strive for energy savings—were more likely to make efforts of the 'relatively painless' "good housekeeping" type'. They did not show evidence of seriously considering long-term energy savings calling for significant investment, product or process innovation, rationalising plant or changing input or product mixes. All this clearly reflected the low relative cost of the energy component in these cases. This is not of course specifically a problem affecting coal but all sources of energy. It indicates a continuing attitude of under-rating the portents of impending energy shortage, which will have to be altered as fuel prices continue to rise.

Understandably the group of energy conservation schemes of the UK Departments of Energy and of Industry introduced in 1977 have included surveys of energy intensive industries. Entitled the Energy Audit Scheme, these surveys have the aim of establishing how much energy is being absorbed in manufacturing processes, in products and in the feed materials and components. The information is due to be used to identify opportunities for reducing energy consumption. Industries to have the benefit of these analyses included iron casting, refractories, petroleum refining, several branches of chemicals, cement, glass and others.

Other organisations are also due to have subsidised help offered so that they can use fuels and material more efficiently. But it is the energy intensive ones that will be the subject of government-funded researches in depth.

An ingenious way of finding where efficiency can be improved is based on directly measuring from aircraft or satellites where heat is being radiated and in what quantity. Scientists at Harwell (UK) are operating such a survey, the results being displayed on a map; temperature differences as small as 0·3°C can be shown and places differentiated to an accuracy of a foot. Industrial heating managers are reported to be interested in the scheme which can show, for example, if buildings, plant out in the open (such as reaction vessels), or pipes are losing excessive amounts of heat.

Finally, the EEC's programme for promoting energy conservation included, in 1977, inviting proposals for R & D on assessing the specific energy consumption of equipment, processes and techniques. So, all in all, it can reasonably be expected that within the next few years we should be both far better informed on the efficiency with which industry is using its energy supplies, and actively in the course of improving it.

Further Reading

Committee on Air Pollution. Interim Report, Cmd 9011, HMSO, 1953

Committee on Air Pollution. Report, Cmd 9322, HMSO, 1954.

'Preventing Bronchitis' Office of Health Economics, 1977.

'Clean Air Today' HMSO, 1974.

Pollution Papers published for the Department of the Environment:
  1. 'Monitoring of the Environment in the UK.' HMSO, 1974.
  4. 'Controlling Pollution.' HMSO, 1974.
  9. 'Pollution Control in Great Britain: How it works.' HMSO, 1976.
  11. 'Environmental Standards: A description of UK practice.' HMSO, 1977.

'Health and Safety. Industrial Air Pollution, 1975' HMSO, 1977.

'National Survey of Air Pollution 1961–71' and annual and other publications of the Warren Spring Laboratory. HMSO.

'The Efficient Use of Energy' edited by I. G. C. Dryden, IPC, 1975 (The definitive (and mammoth) technical review of current best practice).

'National Society for Clean Air Year Book' Annually.

CEGB Annual Reports and Statistical Yearbooks.

'District Heating combined with Electricity Generation in the UK.' Energy Paper No. 20, HMSO, 1977.

Total Energy Conference, 1971 Proceedings. Institute of Fuel

Energy Recovery in Process Plants, 1975 Proceedings. Institution of Mechanical Engineers.

'Energy use in UK industry' by John Chesshire and Christopher Buckley, Energy Policy, p. 237. Sept. 1976.

'A review of gasification for power generation' by B. Robson, Energy Research, Vol. 1, p. 157, 1977.

Energy Audit Series. Department of Energy, 1977.

Annual Reports of the Coal Research Establishment. NCB.

'Fuel Cells: the versatile power source'. IGT Gas Scope, Summer 1977, p. 1. No. 39.

Chapter 6

Fluidization—versatile technique

Solids—we generally learn at school—have some degree of rigidity. Unlike gases and liquids, that can be grouped together as fluids, solids are materials that do not flow. Broadly, this remains true (with some important exceptions) for bulk solids. But solids divided down into particles can be made to flow in a style very like that of a liquid by lifting and agitating them in a stream of liquid or gas. The process is known as 'fluidization'.

Applied to the process of burning coal, it offers the prospect of several improvements, notably of greatly reducing the capital cost of combustion plant. But it has much wider applications. Let us start by looking at the essence of the process.

Suppose we have a 'bed' of small, solid particles, say a few inches deep, resting on a perforated plate and we pass a gas (e.g. air) upwards through this bed at a gradually increasing rate. At low rates of flow the air will more or less leak through without disturbing the particles. But as the flow is speeded up it opens up the bed because the separate particles become in effect suspended in the air stream. The next important stage, as the air movement is accelerated further, is that the bed begins to look and behave like a boiling liquid. You can see that it is now turbulent and agitated, with a very disturbed surface and bubbles rising through it. In this state it flows like a liquid and it has a head of pressure like a hydrostatic head of a true liquid.

Increase the gas flow even further and the bed eventually begins to disintegrate because the particles are blown out of it—'entrained' is the expression—in the gas stream. First the finer particles are carried away, then as flow gets faster, successively larger ones. This obviously implies that for any particular bed there are upper and lower limits of gas flow for satisfactory fluidization essentially of this kind. Of course if the solid is being transferred to a higher level in the course of a process, it can be carried along deliberately in the faster gas stream; and to transfer it to a vessel at a lower level it can be allowed to fall by gravity down a pipeline with an appropriate (lesser) fluidizing flow of gas.

What are the advantages of handling solids in this state?

The intensive mixing means that in any process where heat is generated, removed or exchanged the bed is remarkably uniform in temperature. Since chemical reactions are greatly influenced by temperature and there is sometimes only a small gap between temperatures favouring a desired reaction and those yielding useless unsaleable products, this can have a major effect on the profitability of a process. The bed also has a high heat capacity compared to that of the gas passing through it; consequently if there is any surge of temperature within the gas, the bed tends to stabilize it. Heat is transferred very quickly between the particles in a bed and between them and the gas or liquid passing through them, and to the containing walls. For these reasons, large beds of hot fluidized catalysts have been found to have an overall range of temperature of less than 10°C during test surveys.

Then, since the solid in this state can be handled so much like a liquid, it means that processes previously carried out in successive batches can be converted to continuous operation, which generally favours more precise control of conditions, more uniform quality and cheaper production costs. Handling of solids into and out of the plant where it is due to react is all much more convenient when it is in this flowing state.

Conditions also offer such excellent contact that they are particularly favourable for a wide range of chemical reactions—between solid and gas, between fluidized solids, between fluids (gases or liquids) and solids. And though there are disadvantages—noted later below—the Kirk–Othmer Encyclopedia lists about a hundred organizations and processes in a 'representative sampling of fluid bed applications which have been investigated over the past twenty years' in a list that is described as 'far from exhaustive'. If we exclude applications to coal processing for the moment, the list is seen to include drying solids, coating particles with ceramics, and a very wide range of chemical reactions concluding (alphabetically) with zinc ore roasting.

Probably the most extensive use has been in promoting gas reactions by passing gases through hot fluidized catalysts, notably in the petroleum industry in the cracking processes we discussed in Chapter 2. It is here particularly that full advantage can be taken of the favourable features noted above. The evenness of bed temperature prevents any development of 'hot spots' due to fast reaction localized on a particularly active portion of catalyst. As the catalyst in use becomes coated over with carbon, it is made to flow continuously to another reactor chamber where it is regenerated. So the main operation of cracking can be operated continu-

G

ously too and with accurate automatic controls. The regenerated catalyst of course flows back into the reactor.

Fluidized bed tailings combustor (NCB). Coal washing produces a waste (tailings) containing fine mineral matter, mainly clay, and some combustible solids suspended in water. Millions of tons of tailings are produced in the UK each year. This 1·5 m (5 ft) diameter fluid bed combustor at CRE was designed to prove that they can be burnt. Tailings leave the washery at 95% water. They are concentrated to 50% water in a thickener before being injected into the combustor. The heat from burning the combustible solids evaporates the rest of the water.

Are there any drawbacks?

Some solids are sticky, or tend to agglomerate or for other reasons such as shape of particles do not flow readily. These evidently are not suited to this method of handling; for them other types of process plant or reactors will be preferable. Where the fluidized bed is used for a chemical reaction, the distribution of sizes of the particles may alter as they react. What starts as a dynamically stable bed may produce fines that are carried away by the gas flow, which will mean that appropriate means for trapping and recovering them will be needed in the plant design. There are other detailed technical complications that may add to operating difficulties such as the inevitable reduction in pressure of the gas, thus demanding larger compressors than would be needed for the same operation with the solid in a rotary kiln or a tray with gas passing over it.

The very uniform temperature—often an asset—can prove a disadvantage in some catalytic processes. Sometimes a process works better with a bed operated at two different temperatures in different parts. The simple fluidized bed will not achieve this but there are ingenious techniques with internal baffles that in effect split the bed into two or more zones and it may be possible to design to meet these technical requirements.

Where a counter-current action is needed, a succession of beds may be used. In this pattern of use the solid is arranged to flow in the opposite direction in the series of beds to the gas, although there will not be general counter-current flow within each bed. This again may add to the complication of a process which can otherwise operate in other types of reactors. And in some cases the general relationships of solid size, quantity and properties and of the gas flows in the process simply do not lend themselves to operating by these means.

So, although many industries, and outstandingly that dealing with cracking and reforming of petroleum fractions, have introduced fluidized reactors on a large scale, there are nevertheless many other cases where the balance of advantage does not lie in operating in this way.

How does it work with coal?

Burning coal

Applying this approach to the burning of coal (see Fig. 1)—and incidentally of many other combustible materials—has proved to offer a remarkable range of advantages. Though inevitably there remain technical problems, there are now designs available for generating power, for small boilers, medium size industrial steam plant, for incinerating a wide variety of wastes and for industrial drying. In respect of large-scale plant for raising steam to generate electrical power there seems to be some difference of opinion between the NCB and its associated companies on the one side, and the Central Electricity Generating Board on the other.

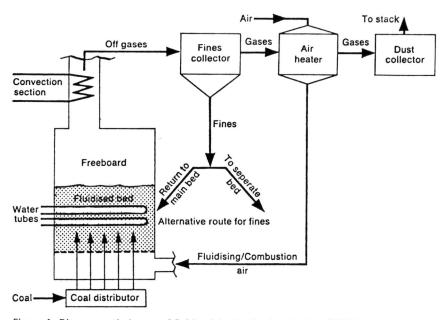

Figure 1. Diagrammatic layout of fluidized bed boiler installation. (NCB).

The latter 'has concluded that they have little immediate potential in the sizes required for its installations', although the early tests had been carried out at the CEGB Marchwood Engineering Laboratories. At that time (some twenty years ago) this line of development was transferred to the NCB as Dr J. Gibson—NCB Scientific Member—has explained 'because of the CEGB's involvement in development programmes for pulverised fuel firing, oil and nuclear power'. Present statements by the CEGB indicate agreement on the value of fluidized bed firing solely for industrial process heat in small-scale plant. For the rest they are generally maintaining a watching brief on developments.

However, the NCB has constantly shown a very high level of confidence in fluidized combustion. In their research laboratories at Leatherhead—the former British Coal Utilisation Research Association premises—they constructed plant on the scale of 3 Megawatts (thermal) for technical appraisals of an industrial scale boiler and a pressurized combustor for generating power in a combined cycle. This both followed and continued alongside earlier work on several smaller experimental units at the Coal Research Establishment (CRE) at Stoke Orchard. The work has aroused international interest and programmes have been sponsored by US government agencies and by the International Energy Agency.

A company called Combustion Systems Ltd (CSL) was formed in

Fluidized bed combustor for corrosion tests. Section: 0·3 m square. Specimens of alloys for boilers are tested in the combustor and those for gas turbines in the off-gas stream outside the combustor.

1974 jointly by the NCB, the National Research Development Corporation and British Petroleum 'to develop, commercialise and co-ordinate fluidised combustion technology and related processes throughout the world'. The following year a further joint company was formed by CSL and Babcock and Wilcox so that equipment could readily be sold with the expertise that was thus being offered. And in the Annual Report for 1976/77 the Board stated that it saw the major new market in the medium term—presumably within the UK itself—as the industrial market for bulk heat where the objective was to develop fluidized bed combustion (FBC) to full commercial application. This would provide at low capital cost high-intensity combustion appliances which would operate automatically and not require specialised coals.

What then is so superior in this method of burning coal to existing practices of burning it on grates or in the turbulence of pulverised fuel furnaces? Briefly the assets can be summarised as

* flexibility in taking a wide variety of fuels;
* high intensity of combustion (as cited from the NCB statement);
* very rapid transfer of heat (combines with the previous feature to reduce capital cost);
* adaptability to systems which trap oxides of sulphur;
* relatively low temperature for combustion, with several favourable consequences.
* suitability for generating hot gas to drive gas turbines;
* readily linked with automatic controls and easy-feed systems.
  Let us deal with each of these features in turn.

Flexibility. For burning coal the bed is made up of coal ash. The depth will depend on the scale of the unit but a great deal of the development has been in beds between half and one metre deep, with a clear space above it in the combustion chamber about 3 metres high to reduce losses of particles. The concentration of coal introduced into the bed of ash is very low. Different workers have reported burning it successfully at levels ranging from about half per cent to two per cent. Consequently the coal is made to burn as separate small pieces each surrounded by hot inert material. In these circumstances it does not matter whether it has strong caking properties or none, if it is of high or low ash, or has a high content of moisture.

In effect this increases the world's coal reserves since it may now be worth while extracting coal that formerly would not burn in existing equipment. It also means that coal can be retrieved for its heating value from dumps of colliery waste or from the high ash material rejected from the so-called washeries where coal from the pit is separated into fractions of different specific gravity, corresponding to different degrees of purity.

Among the strange fuels that have been successfully consumed at CRE have been rejects from colliery washeries with variable high ash and 50 per cent of water, coal of 60 per cent ash, municipal waste (after removing glass and metal) and the coke from the oil of North American tar sands.

High intensity of combustion. For every square metre of cross-sectional area of the bed coal can be burnt at rates up to 400 kilograms per hour—more than double that on conventional coal stokers, although combustion can be maintained at rates of about one tenth of this when the demand for heat is low. CSL state that combustion intensities of up to 50 times those allowable in conventional pulverised-coal furnaces can be obtained on their equipment.

Very rapid transfer of heat. The high rates for heat transfer from the burning coal are achieved by having tubes with the working fluid—normally water/steam, but other fluids are also used—actually within the bed. In boilers this generally results in about half the heat being extracted at this stage, the remainder being transferred to tube systems above the combustion chamber, and from the hot flue gases further in their travels in essentially the usual way.

Adaptability to systems that trap oxides of sulphur. By incorporating limestone (cheap, mainly calcium carbonate, mineral) or dolomite (a natural mixed carbonate mineral of calcium and magnesium) in the bed, sulphur in the fuel can be 'trapped'. It reacts and is fixed in the bed. If the quantities of carbonates are twice those required on grounds of chemical equivalence about 95 per cent of the sulphur is retained, thus reducing the amount discharged into the air as oxides of sulphur.

Relatively low temperature for combustion. Experience has shown the most suitable operating temperature to be in the range 750°C to 950°C as contrasted with temperatures of 1400 to 1600°C in pulverised fuel firing. This eases many problems. If there are sodium salts in the mineral matter of the coal a high temperature of combustion tends to make them evaporate and be carried off in the flue gas (this is known as 'volatilisa-tion'); they are then liable to condense as sticky solids in the relatively cooler tubes of the economisers, where the flue gases are used to heat up the feed water, causing blockages. The lower operating temperatures thus reduce this source of fouling, and also of corrosion of boiler tubes.

In recent years there has been growing attention to the environmental damage caused by oxides of nitrogen. Since these consist of a mixture of compounds they are referred to as NO_x. At the lower bed temperatures, emission of these gases—formed by reaction of the principal constituents of the atmosphere—is also reduced. On the other hand the reactions that cause the oxides of sulphur to be retained are favoured. Both effects work in the direction of protecting the environment. Temperatures are controlled down to the levels required both by the rapid rate of removing heat and/or by supplying excess air above the quantities required for combustion.

Suitability for generating hot gases to drive gas turbines. As we indicated briefly in the previous chapter, it is corrosion or erosion of the turbine blades by the combined chemical and mechanical effects of the impurities swept along in the hot gases from combustion that has hindered the use of coal firing with these prime movers. We have also noted that fluidized combustion reduces the entrainment of alkali salts. Furthermore it results in a non-abrasive ash; and though fine particles do get carried through the dust collecting system they prove to be without effect on the turbine blades. This opens up options of using the hot gases alone for driving a gas turbine—say to generate electric power—or of using combined cycles. In one type of these cycles, the fluidized combustor raises steam for driving steam turbines while the hot gases go on to a gas turbine to generate further power. The Electric Power Research Institute of the USA has commissioned long-period corrosion tests at CRE on a range of alloys that are being considered for gas turbines driven in this way.

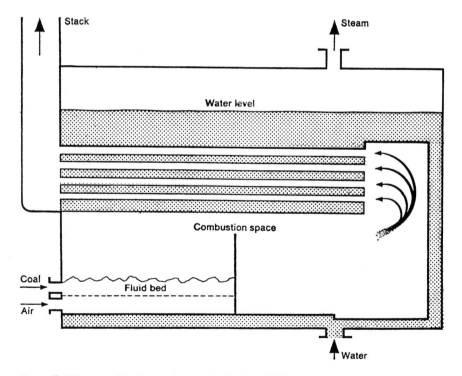

Figure 2. Diagram of bed in horizontal shell boiler (NCB)

The last special feature—that fluidized systems are readily linked with automatic controls and easy-feed arrangements—calls for no further explanation since it follows logically from the fact that the solids flow and fluid controls can readily be adapted to this application.

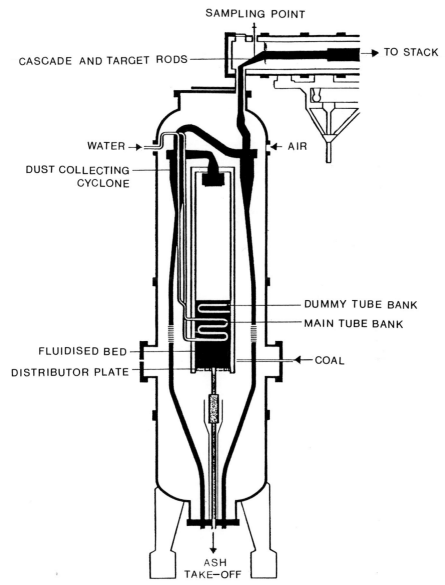

Figure 3. Schematic of pressurized fluidized combustion experimental rig. (Combustion Systems Ltd.).

Because of these advantages, successful uses have included generating hot gases for drying grass on a practical scale on a farm. Elsewhere, in a shell-type boiler for raising steam, the fluidized bed has been used in place of a conventional stoker without tubes immersed so that the approach permits direct conversion of existing boilers (Fig. 2); further variants mean that those now burning oil or gas can also be converted.

Working at higher pressures

If plants are operated at higher pressures there are further gains. Thermal efficiency of plants is improved and most notably the size is reduced for a given duty (Figs 3 and 4). CSL quote that twenty different detailed costings have been carried out by six different organisations including consulting engineers, boiler manufacturers and turbine makers in Britain and the USA and have confirmed savings in costs over current practice. They give these figures:

Cost saving of fluidized combustion power generation compared with conventional practice in the range 120 to 660 MW (e).

	Capital saving	Operational saving
Without sulphur emission control	12 to 20%	10 to 14%
With sulphur emission control	14 to 23%	12 to 16%

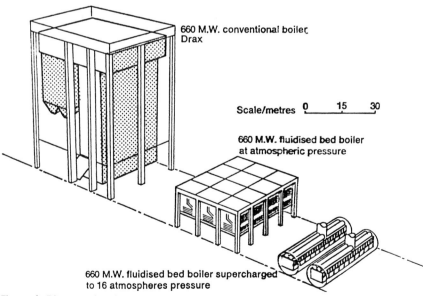

660 M.W. conventional boiler Drax

Scale/metres 0 15 30

660 M.W. fluidised bed boiler at atmospheric pressure

660 M.W. fluidised bed boiler supercharged to 16 atmospheres pressure

Figure 4. Diagram showing comparison of power station boiler sizes. (NCB).

Overall the effect of the lower capital cost and higher efficiency of generating power has been estimated by Dr J. Gibson as likely to yield electricity some 10 to 16 per cent more cheaply than by pulverised fuel firing. Working under pressure is more complex but is considered by designers to show advantage for combining the use of steam and gas

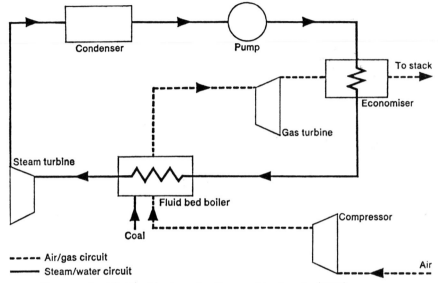

Figure 5. Pressurized fluidized bed combustion combined cycle. (NCB).

turbines (as in Fig. 5) and to be particularly suited to applications for stations of greater than 100 MW(e). And the heat still left in the flue gases even after these operations—technically described as 'sensible heat'— may be recovered for use on the plant or in district heating if conditions justify it.

A technical review by Messrs B. Locke and H. Lunn of the various ways in which fluidized combustion can be applied describes some of the possible working cycles for those who wish to pursue the subject in more detail; it pays particular attention to applying the pressurised versions for generating electric power employing gas turbine cycles. An ingenious development for dealing with peak loads is one linked with storing air— say in an underground cavity—to boost output when needed since normally two-thirds of the power of a gas turbine is used to drive the air compressor. During off-peak periods the generator is used as a motor for driving the compressor and putting the high pressure air into store. Then at the peak periods the whole of the turbine output drives the generator while the stored air is used partly to provide fluidizing/combustion air, partly in tubes immersed in the bed and heated.

In the USA the Energy Research and Development Administration (ERDA) is devoting several tens of millions of dollars a year (the precise amounts have been amended from time to time) 'in developing FBC technology to the point of commercial application'. The parts concerned with working at atmospheric pressure include a so-called 'retrofit' project to demonstrate the feasibility of converting existing power generating systems. Within the group of projects forming the programme for work on pressurised systems are these:

—converting a combustor for municipal waste to a coal-burning prototype;
—a 'modular integrated utility system' for disposing of refuse from a community housing development by FBC delivering electric power and steam for space heating;
—a large-scale combined cycle test facility due to be ready to operate in 1981;
—others to parallel the large-scale project as demonstrations of FBC in applications to industry and institutions, for operational testing in the early eighties.

Wayne A. McCurdy, Chief of the Combustion Division, has stressed in discussing these projects, that they aim to safeguard the environment too; they are placing 'equal emphasis on the control of airborne pollutants and the resolution of technical problems'.

In the UK a power plant being built at Grimethorpe in Yorkshire has a thermal capacity of 80 Mw so that it is large enough to provide lessons for designing commercial-size generating stations. Consuming over 10 tons of coal an hour it is due to provide experimental facilities for research on combustion, heat transfer, cleaning the gas and recovering energy otherwise discharged. The £17 m. project is operated by the NCB on behalf of three countries collaborating in an International Energy Agency (IEA) Agreement; the countries are the USA (Energy Research and Development Administration, ERDA), West Germany (Kernforschungsanlage Jülich) and the UK itself. The NCB has provided the site and £9 million worth of plant is being provided from the participating countries.

However it should be added that the CEGB continues to doubt the value of using fluidized methods in combined cycles. Combining a gas turbine with a steam boiler and turbine or district heating is assessed by them as adding complication and costs, and also adding to the restraints on operation. This remains their view whether the hot gases are produced by combustion in the pressurised bed or by gasification instead of combustion. After examining these possibilities in a recent review, Dr D.

B. Leason, their Generation Studies Engineer, concludes 'it would be unwise to dismiss the fluidized bed completely for power generation but the way is strewn with many difficulties, not least the speed with which the problems can be overcome . . . There is an overwhelming case for steadily improving the existing types of steam plant rather than for changing to novel unproven systems'.

Other applications

Several countries have been working on applying fluidization to pyrolysing coal. Partial combustion can be used to generate enough heat to decompose the rest of the coal feed. In this case, in contrast to the action in the conventional carbonizing process where heat has to pass through a wall to reach the coal, the heat is internally generated yielding variants of the normal products of carbonization. This can readily be used to pyrolyse a coal that will not make a strong coke and thus cannot be handled in a coke oven. The immediate solid product is a 'char' which passes on to the next stage in a briquetting plant where it is compressed without binder to form briquettes that will burn without making smoke. Complete units of this type are now commercially available.

Then, the flexibility in burning a variety of fuels has been found to extend beyond the range already quoted. Coal itself has previously had to be in the form of small particles—either as-received or crushed down to size—in order to be fluidized. But CRE has also developed a way of burning uncrushed coal of sizes up to 50 mm in a process intended for industrial boilers. In this approach the larger coal 'floats' on the fluid bed and burns. Being able to handle large coal extends the scope of this combustion method without the need for crushing; it also reduces the carry-over of fine particles in the gas stream and is found to allow the system to operate with a shallower bed and with reduced height of combustion chamber.

Furthermore, although the above discussion has been mainly devoted to burning coal, FBC can be applied to many other fuels as we have briefly indicated earlier. Using sand as the bed material—and if necessary adding limestone to fix sulphur—a wide range of petroleum fractions can be satisfactorily burnt. When liquids or gases are being burnt in this way, they are fed through a large number of entry pipes distributed over the full cross-section of the bed because they do not spread out evenly over the surface if they are fed in at only one or a few points.

Natural gas can also be handled in a fluidized bed—though the sceptic may well ask what is wrong with using normal burners. Interest in the apparently more complicated method stems from the great advantage of the much higher rates at which heat is transferred—for example in heating products in industrial processes. For this reason the South California Gas Company has commissioned the (US) Institute of Gas Technology to explore this application with particular responsibility for identifying industrial heating needs that might be better met with natural-gas-fired FBC. CSL evidently regard this flexibility in using fuels as being so valuable to users that they offer combustors with multifuel feeding arrangements incorporated in the furnace design. This facility for switching at will from fuel to fuel enables the operator to take advantage of relative fuel cost and availability at any time using the same plant.

But the significance of the flexibility extends beyond this. For it also offers the prospect of linking together plant systems for converting coal but yielding a char, with combustion systems. We shall return to this later but first intend to consider some of the wide variety of approaches now under investigation as possible bases for the future era of transforming coal into materials that are much more valuable.

Further Reading

Article 'Fluidization' in Kirk–Othmer 'Encyclopedia of Chemical Science and Technology'. 1965. John Wiley & Sons Inc.

'Fluidised Bed Combustion' J. Gibson. Coal and Energy Quarterly, No. 7, Winter 1975.

'Fluidised Combustion'. Brochure published by Combustion Systems Ltd, London SW1.

'Clean heat and power cycles using fluidised combustion' by H. B. Locke and H. B. Lunn. The Chemical Engineer, p. 667, Nov, 1975.

'Coal Utilisation Research other than conversion in the USA' by Wayne A. McCurdy. Office of Coal Research, US Department of the Interior.

'Future Power Stations—what will they be like?' by Dr D. B. Leason. CEGB, 1976.

Chapter 7

Chemicals and fuels from coal; the developing future

It is largely the developing shortages of petroleum and of natural gas, and the estimates of large reserves of coal, that have stimulated interest afresh in transforming coal into both chemicals and fluid fuels. In respect of fuels the essence of the chemical problem is summarised in the following short table showing the differences in composition. For coal and oil these must naturally be given as ranges, but methane—usually the main component and sometimes the only one of natural gas—is a chemical compound of fixed composition.

Fuel	Carbon	Hydrogen	Oxygen	Nitrogen	Sulphur	Ratio H/C
	%	%	%	%	%	
Bituminous coal	75–92	5·6–4·0	20–3	2·0–0·75	up to 1	0·057
Crude oil	84–88	14·5–11·5			up to 5	0·15
Fuel oil	about 85	about 11			up to 4	0·13
Methane	75	25	0	0	0	0·33

Note: where compositions are given as ranges, the ratio has been calculated from the mid-points.

Clearly the ratio of hydrogen to carbon is much higher in the liquids and higher still in the gaseous fuel than in the coal. Consequently converting coal into the other products means introducing more hydrogen and/or separating off compounds with a higher proportion of hydrogen leaving behind a carbon-rich residue. The hydrogen may be supplied as such or in the form of steam. The steam is obviously cheaper but its reaction with carbon absorbs heat (it is 'endothermic'); in turn this means that heat must be generated in the reactor in order to sustain the reaction

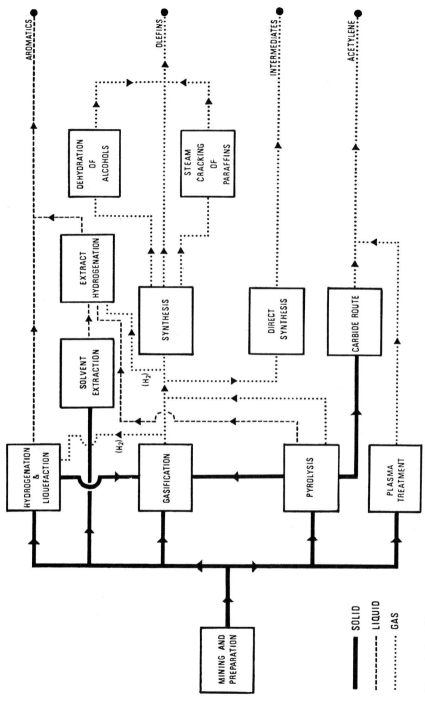

Figure 1. Possible routes to chemicals from coal. (From 'Various Aspects of the Use of Coal in the Chemical Industry' — a paper prepared by Dr. J. A. Lewis of the United Kingdom Department of Energy on behalf of the Coal Committee of the United Nations Economic Commission for Europe, Geneva, 14th June 1976).

between coal and steam. It may be provided by supplying oxygen, or by air which is cheaper but dilutes the resultant gases with its nitrogen. It is attempts to resolve the dilemmas indicated by these features, and to raise the thermal efficiency of conversion that tends to account for the wide variety of processes now under development. Often the processes may be adapted to produce materials that can be sold as 'chemicals' rather than fuels, or may yield both types of product simultaneously, and the following discussion deals with coal conversion with both objectives.

Though methods may be grouped together in various ways—as indicated in Fig. 1 where the emphasis is on reviewing possible routes to chemicals—the essential technologies can reasonably be classified under these headings:

A. Separating 'hydrogen-rich' compounds

1. Pyrolysis.

When coal is heated in the absence of air the yield of volatile products formed and driven off ranges from almost 20 per cent in existing low-temperature carbonization to about 25 per cent in coke-ovens; applying newer methods, this proportion can be considerably increased.

2. Extraction with liquid solvents.

If coal is treated with anthracene oil at about 500°C almost 90 per cent of it can be removed in the liquid leaving behind a solid residue of all the mineral matter with the insoluble carbonaceous material.

3. Extraction with a 'super-critical' gas.

Coal has also been extracted using gases at very high pressures and medium temperatures, under conditions called 'super-critical'. In this usage, the term 'critical' applied to a gas means the temperature above which it cannot be liquefied by pressure. At the critical temperature itself, the gas can still be liquefied by pressure and the pressure required is known as the 'critical pressure'. In the extraction technique both pressure and temperature used are above the critical values.

B. Introducing additional hydrogen

1. Hydrogenation

Research projects on adding hydrogen include processes where it is

H

applied under pressure and sometimes in multiple stages to

* coal itself mixed into a slurry with various oils, usually derived from coal;
* coal at the same time as pyrolysis, when the method is known as hydropyrolysis;
* coal at the same time as gas extraction to increase the yield of liquid products;
* pyrolysis liquids to produce 'lighter', more volatile and more valuable fractions;
* the material from extraction with liquids to yield lighter fractions;
* the material from extraction with gases to yield lighter fractions;
* residues of high molecular weight which are 'cracked', when the method is known as 'hydrocracking'.

2. Gasification

When 'gasification' is used for adding the hydrogen—that is by reacting the coal with steam and air or oxygen and converting it into gas—mineral matter and sulphur compounds are removed from the gas stream. Dependent on the conditions used, the gases may contain methane, carbon dioxide, carbon monoxide, hydrogen and nitrogen. Further treatments may be designed to result largely in fuel gases, alcohols or the range of products of the Fischer–Tropsch process. If the gases are intended for fuel gas, they may be used as low or medium heating value fuel (in the familiar old units 100 up to 550 Btu per cubic foot) or upgraded to a fuel rich in methane (up to 1000 Btu/cu ft) usually called Substitute Natural Gas, SNG.

The coal may be converted into gas under pressure—of advantage for later gas reaction stages but involving engineering difficulties in handling solids in and out of pressure vessels—or not; and the reaction may take place in a so-called fixed bed—actually a slowly descending one—or a fluidized one, or may occur co-current with the gas stream.

An alternative grouping of the technologies is into those concerned primarily to produce liquids, obviously known as 'liquefaction', in contrast to those where the primary interest is in the gases where the term used is gasification, as above. Though this may seem obvious, processes may produce both, or gases intended for further reaction to liquids, and also a solid, so that the naming tends to indicate a main objective rather than a single class of product.

Yet a further approach to conversion is based much more on the fundamental chemistry of what happens. This approach treats the conversion as being a chemical degradation or alternatively a synthesis. In this sense, the term 'degradation' means a partial breaking down of chemical structure. Of the technologies named above, pyrolysis, with or without the action of hydrogen, with or without extraction by some solvents ('degrading' solvents, phenanthrene being an example) degrades the complex

Experimental coal gasification rig. This rig is for the gasification of coal in air or air/steam mixture at temperatures up to 1050°C and pressures to 6 bar. The reactor, 0·1 m diameter, 1·2 m high, is contained within the brickwork showing below the platform. (NCB)

structures of the coal into much simpler ones. The other category—synthesis—also depends on the most thorough breakdown first to the very simple mixture of gases already noted under the heading of gasification, but refers to what follows; if they are used for chemical purposes, the mixture after purification is evidently a 'synthesis gas' and this expression is particularly applied to the mixture 'CO + hydrogen.'

Degradation tends to yield mainly aromatic compounds, while those emerging from synthesis are almost entirely aliphatic. All of the processes are likely to be considerably influenced by relatively small quantities of materials which act as catalysts, that is substances that affect the rate of chemical reaction without being consumed—or in some cases are thought to be consumed and then regenerated. Certain types of mineral matter naturally present may also act in this way. Let us expand a little on the work in progress under the headings of the various groupings of technologies.

Pyrolysis

Some of the developments in pyrolysis processes were briefly mentioned in Chapter 4. Dr J. A. Lewis of the UK Department of Energy has summarised how they compare with traditional coke ovens in the following table:

Operating conditions and yields for developing pyrolysis processes.

Process	Coke oven	Lurgi-Ruhrgas	COED	Garrett	CSIRO
Operating temperature deg C	900	650	450–800	600	700
Residence time (order of magnitude)	Hours	Mins	Mins	Secs	Secs
Process yields in weight per cent					
Char residue or coke	75	67	54	52	40
Tar and oil	5	16	18	32	17
Gas	16	4	14	6	34
Aqueous	4	13	4	10	9

Of these the coke oven techniques have been discussed; the Lurgi–Ruhrgas is a rapid coking technique being applied to lignite in plants operating on 1600 tons a day in Germany. The COED is a multi-stage group of units where coal is pyrolysed in fluid beds at successively higher temperatures. Both this and the Garrett are American. The latter entrains pulverised coal into a reactor and rapidly cools the products to prevent any further chemical changes. The initials CSIRO refer to the Australian Commonwealth Scientific and Industrial Research Organisation, where a flash pyrolysis process using no hydrogen is operating so far only in laboratory units. A large pilot plant is scheduled for 1980. Even higher temperatures and shorter reaction periods are achieved in the electric plasma arc which is under investigation in a number of countries. The work at CRE includes pyrolysing coal (or feedstock derived from coal) in a direct current jet operating at 30 kW at temperatures above 1300°C to make acetylene; pyrolysing coal itself in a plasma arc in compressed hydrogen is favourable to forming aromatics. Temperatures up to 10,000°C have been reported, applied for a few hundredths of a second and resulting in yields of 50 per cent of gas, but on a small scale.

Extracting with liquid solvents

This approach is often being combined with hydrogenation but not inevitably. In the USA the pressure for 'clean' fuels—that is those causing minimal damage to the environment—has led, among other avenues of approach, to using solvent extraction to yield low-ash, low-sulphur fuels for boilers. In this area of development they nevertheless are looking to the possibilities of products for the higher value markets of petrol and chemical feedstocks. They have a plant now operating on 50 tons a day of coal and a design for a 250 ton/day plant using what are called 'donor solvents', that is those that both dissolve a portion of the coal and transfer hydrogen (such as tetrahydronaphthalene, known as tetralin).

CRE are using extraction with anthracene oil to make very pure carbon from coal. The solution is separated from insoluble residue, solvent recovered and the extract heat-treated in stages, including delayed coking. After the solid has been treated at temperatures up to 2700°C the product is a high grade electrode coke suitable for use in arc steel furnaces.

Extracting with super-critical gas

This is again an area where CRE is active and the extraction may again be combined with hydrogenation. A very wide range of chemical types of compounds are recovered—including straight chain aliphatics, isoprene derivatives, hydro-aromatics as well as aromatics—suggesting that they have been separated from the coal in structures more or less the same as those that exist in the coal itself. This is in marked contrast to the products emerging from any other form of treatment of coal (all other treatments

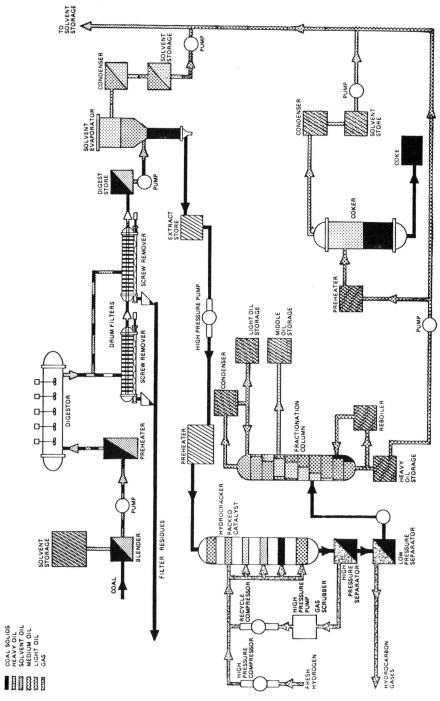

Figure 2. Coal conversion scheme designed at the NCB Coal Research Establishment (CRE) starting with the solvent extraction of coal, to maximise the yield of premium materials. Two main products are envisaged: low-boiling aromatic hydrocarbons, and carbon (coke) probably for the manufacture of electrodes.

break down the coal structure). An interesting derivative is that the process has also been shown to extract oil from oil shale very effectively.

Hydrogenation

Both the liquid, and the critical gas, solvent processes have been applied combined with hydrogenation. Solutions of coal in heavy tar oil are thus converted to low-boiling hydrocarbon oils (in the presence of catalysts). CRE have designed a full conversion scheme based on this approach, to make both the valuable oils and the ultra-pure electrode carbon (Fig. 2). When hydrogen is introduced into the gas extraction, the solution forms a source of benzene, toluene and xylene hydrocarbons. The difficulties in pigeon-holing processes are well illustrated by the HYGAS process adopted by ERDA as a design study for a demonstration plant (Fig. 3).

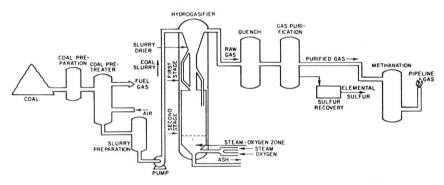

Figure 3. The HYGAS Process (Developed by the US Institute of Gas Technology) Coal is prepared by drying and crushing. With caking coals, 10 per cent is part-oxidized (pre-treated) to render it non-caking. Mixed with light oil from the process it is pumped into the hydrogasifier on to a fluid drying bed. Passing down the vertical reactor of the first stage it reacts with hydrogen from the second stage and hydro-generator below. The residue reacts at higher temperature with hydrogen in the second stage. Two thirds of the methane is formed directly in these reactors accounting for the high efficiency of this process. The remaining char falls into the bottom zone for converting by steam and oxygen into hydrogen and oxides of carbon.

For this is in essence based on fluid-bed hydrogenation of coal. But the first stage is based on entraining the coal before it passes to two successive fluidised reactors for hydrogen treatment; furthermore the hydrogen comes from a further reactor where char is gasified in steam and oxygen (although other methods of generating the hydrogen are also being considered). In this the emphasis is on producing a gas of high calorific value, that is with a high content of methane, variously known as SNG and a 'High Btu gas'. Other processes among the many supported by

ERDA are directed much more to yielding a synthetic equivalent to a crude oil. For example in those known as H-COAL and a further one called SYNTHOIL coal is made into a slurry with an oil derived from coal and treated with hydrogen; associated programmes assess the resulting liquids as sources of petrochemicals.

Gasification

It might appear that the key interest here would be to form SNG at about 1000 Btu per cubic foot—and there is in fact a considerable US effort in this area to produce gas free of sulphur and other pollutants at high pressure, burning in the same way as true natural gas so that it can simply be fed into pipelines in place of the natural product. But several countries—including the US—also see the value of generating a 'Low Btu gas', with heating value between 100 and 500 Btu per cubic foot. For this calls for a less complex plant; both capital and operating costs are lower per Btu generated. And provided the gas is made close to the point of use, it has a good potential market in both industry and electric power plant. The combined-cycle test plant at Pekin, Illinois (mentioned in Chapter 5) is designed to test this concept in an integrated system making a low Btu gas, purifying it and feeding it to gas turbines generating electricity. In both this and the other gasification programmes there is great emphasis on protecting the environment, and associated researches are wholly devoted to this aspect.

Turning back to High Btu gas, a further demonstration plant of ERDA is due to use an important development of the Lurgi gasifier. In the long-established version of this plant pre-heated steam and oxygen at pressures of 20 to 30 atmospheres react with crushed, dried coal in a slowly descending bed at temperatures around 900°C. Temperature has to be kept down to this level to prevent the ash melting, and excess steam has to be used for this purpose. This reduces thermal efficiency and has other disadvantages. To overcome them, Dr D. Hebden (Programme Director at the Westfield Centre of British Gas) picked up the threads of earlier work in eliminating this excess steam. Financed by an American group, a programme of trials was developed using oxygen with only the amount of steam required for gasification, allowing the temperature to rise so that the ash melted and was run off from a re-designed bottom as a liquid slag. Tested on a range of US coals, results have been successful. The slagging gasifier produces gas at several times the previous rate, at higher thermal efficiency. The gas is then cleared of tar, partly reacted with steam in the 'shift reaction' transforming CO into CO_2 and hydrogen, then CO_2 and other acid gases removed. This brings the composition to the correct ratio for methane synthesis according to the reaction $CO + 3H_2 = CH_4 + H_2O$.

Cooling removes the water leaving a gas containing only methane. This combined German, British and American development of a first generation gasifier is due to be run by the Conoco Co. on a scale using 3,800 tons of coal a day (Fig. 4).

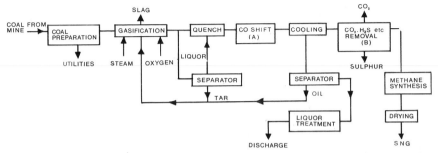

Figure 4. The production of SNG from coal by the Slagging Gasification route. (British Gas) A is the CO shift stage converting CO to CO_2. B is the Rectisol process for removing acid gases and other impurities.

Other gasifying systems being investigated are called 'second' and even 'third' generation, implying rather longer time-scales. Interest in them arises because the established techniques are expensive, thermally inefficient and often limited in the kinds and sizes of coal that can be handled. Among those now being tested is the unique CO_2-acceptor (Fig. 5). In this there are two fluid-bed reactors, a gasifier and a regenerator. Suitable coal is fed to the gasifier where it quickly loses its volatile matter and is gasified with steam, at about 800°C and 10 atm. pressure. The heat is

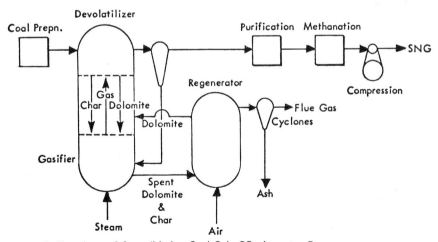

Figure 5. Flowsheet of Consolidation Coal Co's CO_2 Acceptor Process.

supplied by a circulating hot stream of lime-bearing material. But most of this heat is derived from the chemical reaction of CO_2 (formed during gasifying) with the lime. Hence the name CO_2-acceptor. The carbonated material—it may be limestone or dolomite originally—passes back to the regenerator. Here it is heated by burning char from the gasifier, in air, and the CO_2 is driven off from the carbonates decomposed at about 1000°C. The flue gases from the regenerator, containing nitrogen and CO_2, are separate from the gas formed in the gasifier. So the process yields a synthesis gas suitable for converting to SNG without the expense of oxygen; its composition also eliminates the need for a CO shift convertor and—being almost free of the acid gases CO_2 and H_2S—needs little 'clean-up'.

ERDA'S extensive later-generation programme also comprises the BI-GAS slagging gasifier. Unlike the Lurgi development this is an entrained coal system, in two stages, using oxygen and steam, and operating at higher temperature on all ranks of coal. No tars or oils are formed and the objective is conversion to SNG.

In this programme too are the SYNTHANE and the HYDRANE processes, and one called 'Self-Agglomerating Burner Process'.

The main characteristics of the SYNTHANE are fluidized bed pre-treatment at 425°C in oxygen/steam to destroy caking properties, followed by gasifying in oxygen/steam at almost 1000°C at 68 atmospheres pressure; volatile matter is separated and the gas purified then methanated to SNG.

In the HYDRANE system, there is gasification of char with oxygen/steam to yield hydrogen; this then reacts with coal fed into a fluidized bed yielding a high-methane gas directly at high thermal efficiency under high pressure. Very little further conversion is needed to form SNG.

Two stages are applied in the Self-Agglomerating Burner—one for generating heat, the other for gasifying, the gases from the two being kept separate. The name derives from using a fluid bed for burning part of the coal under such conditions that the ash agglomerates into free-flowing balls. This 'collects up' the fine ash and leaves a flue gas that is clean enough to be used in a turbine. The small particles are circulated to the gasifier to supply the heat for the endothermic reaction of the steam with the carbon without needing oxygen.

Underground gasification

Could we make use of the energy and chemical potential of coal without digging it up out of the ground? If coal within its seams could be turned into gas and piped to the surface, its resources would be available without the expense and hazards of sending men underground to hack it out. And this would also save a large part of the further technical efforts we have discussed in earlier pages.

It is a dream that has been shared across national and political boundaries for more than a century. Sir William Siemens spoke of it in Britain as far back as 1868; the Russian chemist Mendeleyev picked up the idea and major trials have been carried out over many years in the USSR. But the first experiments were in Britain—by Sir William Ramsay in County Durham in 1912. Several other countries have shared in the technical excitement—and the interim failure. There have been projects in the USA, Belgium, Czechoslovakia, Italy, Poland and by the French in Morocco. There are several methods, varying in detail; in all cases the coal underground must be exposed, by drilling down to it or if it is outcropping on to the surface, into it. Then an outlet must be made for the gases that are to be produced. Finally a linking path is needed between the two or more original holes through which gases can flow (Fig. 6a). The

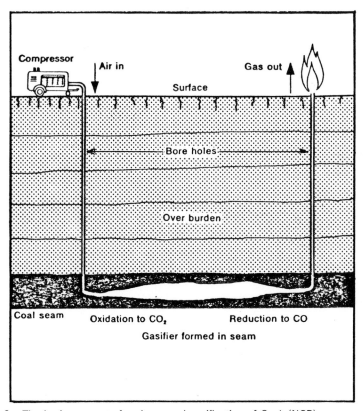

Figure 6a. The basic concept of underground gasification of Coal. (NCB).

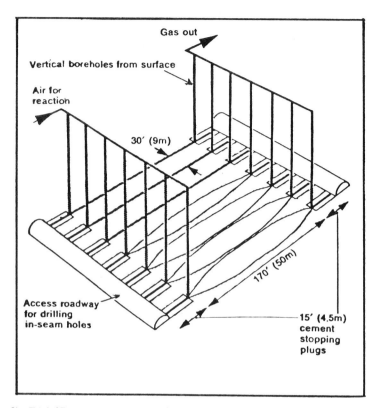

Figure 6b. Trial 45 at Bayton, England (c. 1955) (NCB).

next step is to set the coal alight, by an incendiary device or electrically, and blow air down one hole through the fire and along the linking path. The resulting gases are forced up the other hole. If all goes well, controlled combustion underground can result in a mixture of pyrolysis gases and what is called 'producer gas' (CO + nitrogen). If the burning is not properly controlled all the fuel gas is burnt and only dead flue gas comes up the pipes.

Early researchers lost interest when they encountered technical difficulties in controlling conditions underground. The gas they made was found to be of low calorific value and expensive. It proved very difficult to continue to operate for long periods consistently. Nevertheless ERDA is now operating a site at Laramie, the Russians have three sites working and renewed research has been reported from Belgium, Germany and Canada. Spurring on the interest in underground gasification (UG), notably in the USA have been the new laws tightening up safety and health requirements for miners and difficulty in recruiting the men. These are extra inducements over and above the general drive to find economic ways of converting coal into SNG in the face of current and growing shortages of natural gas.

Yet, simple though the idea is in principle, it continues to be difficult to work in practice. Bear in mind that whatever happens is going on underground, out of sight and can be influenced only indirectly. What is actually taking place in the fire-ravaged coal seam can only be deduced at the time from readings of instruments passed down into the seam and from the quality of the gas. Working in these inaccessible conditions, the research engineers have experimented with such features as using high air pressures, and adding steam and oxygen to gasify underground by the same reactions as in plant overground; associated projects include design of the power plant to use the low Btu gas.

A recent NCB review concludes

'A number of advances in technique would be necessary to make the process a reliable one capable of consistently producing a gas of reasonable quality (about 100 Btu/cu ft, $3 \cdot 7$ MJ/cu m) on a commercial scale over a period of approximately 20 years.'

Key areas picked out as needing successful technical developments were named under no fewer then 12 headings. Among them were access to the seam, linkage, orderly progression of the reaction, gas leakage, faulted areas, preventing the roof caving in, and dealing with a succession of seams one below the other. These are some of the problems underground. On the surface, the gas will need cleaning and may need design of new special plant to use a gas much leaner than those customarily applied. Though there are evidently common features in the difficulties to be met in all the coal producing countries, there are sufficient differences in seams and in coal properties (for example physical permeability) to make it essential to run the trials within any country that wants to apply UG. Mr Leslie Grainger, then Scientific Member of the NCB, estimated in 1977 that a pilot scheme for Britain would cost up to £15 million. And—as a more distant vision—there are tentative ideas for applying solvent extraction and bacterial breakdown methods to coal still in the ground.

Building up chemicals

If the gases resulting from gasification are not being used directly, or after methanation, as fuels, they can be applied for synthesis, or even divided and used in both ways. A most general diagram of possibilities (Fig. 7) shows the usual constituents of the raw gas. They are methane (CH_4), carbon monoxide (CO), hydrogen (H_2) and carbon dioxide (CO_2) with

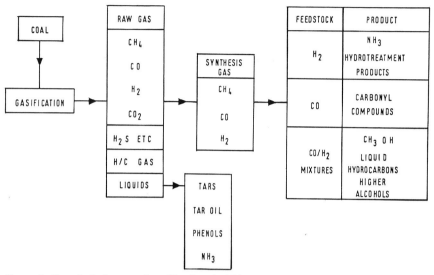

Figure 7. Chemicals from coal gasification. (NCB).

hydrogen sulphide (H_2S) and other sulphur compounds; dependent on the process, there may also be some hydrocarbon gases (shown as H/C) and liquids, modifications of usual coke-oven liquids, that will be separated out. The purification stages will take out the sulphur compounds and some or all of the CO_2 according to requirements. If the principal aim is to make chemicals the remaining gases form a valuable synthesis mixture.

For making methanol (for solvent, further synthesis or for formalde-hyde used in plastics) a mixture of both CO and CO_2 with hydrogen—naturally in the right proportions—is needed. And there is now great interest in this as a route to petrol by a catalytic method developed by Mobil. For a wide range of other reactions, the CO_2 is removed leaving the mixture of CO and H_2 earlier cited as 'synthesis gas'. It is this mixture that is the basis of the SASOL process. Chemically, the essence of this—the Fischer–Tropsch synthesis—is the reaction

$$CO + H_2 \rightarrow -CH_2- + H_2O$$

the symbols —CH_2— representing units or radicals that can link up to form hydrocarbons having carbon chains of various lengths. The Synthol variant used in South Africa with an iron-based entrained catalyst yields lighter hydrocarbons, producing a mixture with a large proportion usable as petrol.

Mixed with olefins (the unsaturated hydrocarbons, the ethylene series), 'synthesis gas' passed over cobalt catalysts forms aldehydes, alcohols and other compounds containing oxygen in what is called the 'OXO synthesis' from a German name; from some of these compounds are made industrially important solvents, herbicides, plasticisers for giving paint flexibility, as well as lubricants.

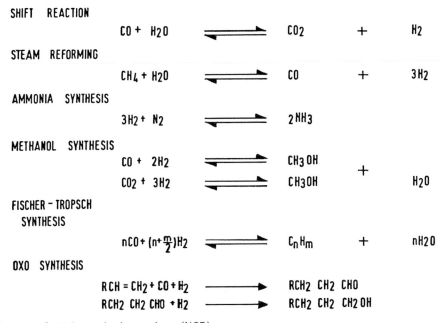

SHIFT REACTION

$$CO + H_2O \rightleftharpoons CO_2 + H_2$$

STEAM REFORMING

$$CH_4 + H_2O \rightleftharpoons CO + 3H_2$$

AMMONIA SYNTHESIS

$$3H_2 + N_2 \rightleftharpoons 2NH_3$$

METHANOL SYNTHESIS

$$CO + 2H_2 \rightleftharpoons CH_3OH$$
$$CO_2 + 3H_2 \rightleftharpoons CH_3OH + H_2O$$

FISCHER - TROPSCH SYNTHESIS

$$nCO + (n + \tfrac{m}{2})H_2 \rightleftharpoons C_nH_m + nH_2O$$

OXO SYNTHESIS

$$RCH = CH_2 + CO + H_2 \longrightarrow RCH_2\ CH_2\ CHO$$
$$RCH_2\ CH_2\ CHO + H_2 \longrightarrow RCH_2\ CH_2\ CH_2OH$$

Figure 8. Main synthesis reactions. (NCB).

But the constituent separate gases may themselves be isolated. CO treated with steam in the 'shift reaction' is converted to CO_2 so that it can be readily removed; the hydrogen after further purification, may go for any industrial purpose but probably the greatest interest is for making ammonia, used in millions of tons throughout the world. Alternatively, major interest may be directed to the CO which can also be separated pure if needed—for this compound is also a valuable feedstock. From it can be made ketones, aldehydes, glycols (all with many industrial uses) but also compounds with metals, called carbonyls, used as steps to making powdered metals of high purity.

The essential reactions are shown in Fig. 8.

Basic science is not neglected

Though it is impossible here to attempt to summarise the wide-ranging programmes on the basic science of coal all the industrialised coal producing countries have shown that they are aware of the need for more understanding of the constitution of coal and of its reactions as essential background for the practical work on technologies for converting it. Again the largest investments are in the USA, though important contributions also come from elsewhere.

Consider for example, carbonisation, the oldest process for converting coal, yet today still a subject of intensive study in respect of such aspects as differences in behaviour of different macerals, classification by microscopy, study of rheology (deformation and flow) as coal changes into plastic material and then coke, relating the structure of coal (from X-ray analysis) to that of the coke. Dr H. Jüntgen of Bergbau-Forschung has quoted research programmes on the kinetics (rates of change) as volatiles are driven off from coal, mathematical analyses of gas formation, on pyrolysing 'model substances'—that is chemicals of simpler known structure as possible guides to what happens in the more complex, less-precisely-known structures of coal. Other lines have included laser-heating, and looking at the inter-play when more than one process is in operation—such as pyrolysis and combustion, or pyrolysis and gasification.

There are also projects in fundamental science as direct back-up to developing processes. A notable case is the work on catalysts, striving to replace trial and error with some degree of understanding of the mechanisms involved, so that the right materials in the right physical form can be chosen for speeding up reactions—an important part of the clue to reducing conversion costs to economic levels. This, in turn, leads us in to the key question below.

Does it pay to convert coal?

If we exclude improvements in carbonisation, we have to face the fact that coal conversion to other fuels and to chemicals—with one important exception—tends to result in products that cannot compete with those already on the market, made by other means from other feedstocks, normally petroleum-based. The ERDA programme for Financial Year 1978, acknowledging this as a fact of life, explains that its great efforts in the area are based on the coal reserves estimated to last several centuries and the potential for achieving energy independence in the USA. But among their multiple objectives is that of 'validating the economic acceptability' of second-generation coal processes. This may well be taken to mean making them competitive with processes based on other feedstocks, at the various target times. That is the purpose of sharing costs and risk of major projects, as Demonstration Plants, with industrial partners to stimulate private industry into proceeding with plants on their own initiative.

For gasification, they quote 'near-commercial scale in the early 1980s'. For liquefaction the support of commercial-scale implementation is for the year 1990, with a further generation of commercial-scale plants by 2000. Advanced power systems—high-performance combined-cycle turbine systems with improved turbine technologies—are given practical target dates as the range 1985 to 2000.

The 'exception' I cited above in respect of current competitiveness is in South Africa. For Sasol II—based on coal measures owned by the same company—the plant was costed assuming that high petroleum prices would continue and coal would be available at prices up to $5 a tonne. This coal price is of a totally different order to those in the other industrialised countries. Compare it with the average income per ton for the NCB in 1976/77 of £19.59 (about $35) and coal prices within the EEC ranging widely between $45 and $90 according to grade and quality. Mr P. E. Rousseau, Chairman of Sasol, reckoned in 1975 that if their coal went up to $10–15 a tonne at the pitmouth, oil from coal would become a doubtful operation economically. Though social and political comment is no part of this book, it is very relevant to refer to continuing disturbances in South Africa in 1976/77 which may well result in higher wages, therefore in higher coal costs, in turn upsetting the financial basis on which Sasol II was founded.

In the UK, ICI have understandably given attention to the economics of possible feedstocks for synthesis gas. They use natural gas as a feedstock in the UK for making ammonia and methanol. Late in 1975, their assessment of the break-even relationships for coal and oil prices was that shown in the graph Fig. 9. This shows for example that at that time, for generating electricity, coal at 7p/therm (about £16 a tonne) broke even with oil at $10 a barrel. On the graph if you plot a point for the price of coal and that of oil, and the point falls to the right of a break-even line, that indicates that oil is the more economic feedstock. Though the absolute figures will naturally be different at other times, so that the lines can only be taken as broadly indicative, the study shows how far out is any prospect of making chemicals from coal by even an improved Fischer–Tropsch process in any of the other industrialised countries today. Considerable movements in relative prices will be essential—but it should be added, that they are expected.

The perspective offered by Mr L. Grainger (then Scientific Member of the NCB) and Dr P. Paul in a paper in 1977 was this:

> The economics of coal conversion to SNG and oil substitutes are still unfavourable, particularly in the UK. When natural gas and oil

I

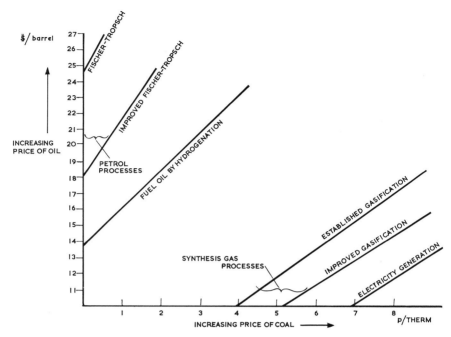

Figure 9. Break-even relationships between coal- and oil-based processes. (ICI).

prices begin to reflect supply limitations, relativities will probably change. Coal conversion processes must be available for commercial operation by that time. It is possible that an international trade in coal conversion products could be established before the year 2000.

The two big coal producers of Eastern Europe have given indications of being more optimistic. In papers to the UN Economic Commission for Europe, the Soviet Union reported

Technical and economic calculations carried out for a hyarogenation plant processing 12·7 million tons of raw material a year (4·2 million tons of coal and 8·5 million tons of oil) and an oil-processing plant with an output of 12 million tons of oil a year have shown that these two methods of producing liquid fuels are comparable as to quality, output and production cost.

This is based on 'a technique of hydrogenation of brown and hard coal together with petroleum product for the production of low-sulphur boiler fuel, motor fuel and chemicals being developed at the Institute of Mineral Fuels.'

For Poland, a report referred to a coal-based chemical industry producing gases for chemical synthesis, gaseous and liquid fuels and other products. Under their 'maximum' programme, the output of the coal-based chemical industry was given as these estimated quantities:

Product	Quantity in 1990
High calorie gas	8,000 million cu m a year
Liquid fuels	5 million tons a year
Coke type fuels	4·3 million tons a year
Energy from power stations using carbonized fuels	24,000 million kWh per year

This output is designed to meet expected changes in the structure of consuming fuels. For they too expect users to prefer their energy sources as electricity, and as gases and liquids rather than solids—while they have large reserves of both bituminous (hard) and brown coal. Hence the interest in conversion. But there is no reference to comparative costs.

Thus, overall it is clear that (Sasol apart) synthetic fuels and chemicals from coal are currently much dearer than those from petroleum in the free market economies. World-wide interest stems from forward-looking policies—as soon as possible to reduce dependence on petroleum and gas imports from politically unstable regions, and long-term, to anticipate exhaustion of petroleum and gas deposits.

Yet there are further developments that seem to offer prospects of sharing operating costs—though adding to the complications—by combining units that are chemically converting part of the coal, with others that burn the residue for some productive purpose. Some of the ideas along these lines are examined in the following chapter.

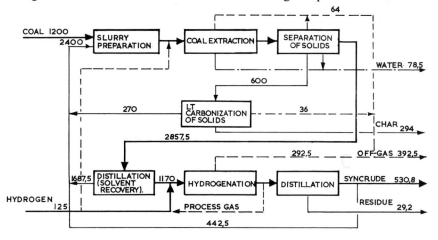

Figure 10. Scheme for research installation for coal liquefaction. (Poland) Figures indicate flows in kilograms per day. (From paper to UN Economic Commission for Europe).

Further Reading

'A national plan for energy research, development and demonstration: creating energy choices for the future, 1976. Vol. 1: The Plan.' ERDA 76–1. USA.

National Coal Board. Coal Research Establishment. Annual Reports.

'Automotive Fuels derived from Coal.' L. Grainger; I.Mech.E. symposium, 1977.

The Robens Coal Science Lecture, 1975: 'The coal renaissance, a South African point of view.' by P. E. Rousseau. Journal of the Institute of Fuel pp 167–175 December 1975.

'Clean Fuels from Coal.' Symposium II papers. Sponsored by Institute of Gas Technology, Chicago, Illinois, USA.

'Coal as a fuel or as a source of chemicals—the new technologies.' Paper by P. V. Youle. World Coal Conference, London, 1975.

Proposals for National Programme of Development on Coal Conversion Processes. PADB 73/17. National Coal Board, 1974.

Coal Industry Examination. Final Report 1974. Department of Energy.

'The Constitution of Coal and its Relevance to Coal Conversion Processes.' J. Gibson. The Robens Coal Science lecture, 1977.

'Underground Gasification of Coal.' P. N. Thompson, J. E. Mann and F. Williams. NCB. 1976.

'Fossil Energy Research Program of the Energy Research and Development Administration. FY 1978.' ERDA 77–33, April 1977, ERDA.

Papers presented to UN Economic Commission for Europe, Coal Committee, June 1976.

'Coal-burn it or convert it?' Leslie Grainger and Peter Paul; I.Chem.E. symposium, 1977.

Chapter 8

Coalplexes—making the maximum use of coal

'In parallel with this programme of plant and unit process development' ran the introduction to the NCB's proposals in 1974 for a national programme of development on converting coal

'. . . . there is a need for analytical studies of the developing energy scene, to study how the separate processes can be integrated and co-ordinated together in various ways as coalplexes designed to make maximum use of the fuel resources available to the nation with due regard to the environmental and conservation aspects.'

Noting that in the past energy industries had developed independently in competition with each other, the document went on to argue, in effect, that we were now in a new situation. This would demand the closest co-operation between energy industries in the future to ensure that energy was used to the highest efficiency, with a minimum of capital investment and a minimum wastage of scarce resources.

Brave words. But are these practical concepts, and how far have they got in studies or working plant? For many years there has been interest in a more limited sense in linking together generating electricity and heating; the waste, as low temperature discharge heat, or over 60 per cent of the energy fed to power stations has long galled engineers. Results of assessing these possibilities have been given in Chapter 5; in the UK the greatest successes have been achieved where a power plant has been installed on the site of a big power user (one or more factories, a hospital or other large institution) and operated in such a way that the discharge heat is put to good use. Using discharge heat for district heating was much more questionable. But the Advisory Council on Research and Development (ACORD) in its discussion document reckoned there were in industry about 300 combined heat and power schemes (CHP) in the UK meeting 20 per cent of industry's demand for electricity. They accepted the 'well-established' case for installing these schemes in industry. Naturally, such CHP schemes may use any fuel and they are quoted here as linking together energy uses, not particularly as applications for coal alone.

However, it is an obvious extension of the principle to examine how far integration can be carried. If the fuel is coal, integration could combine in the same plant such features as

* processing the coal;
* working up the products yielding gas and liquid fuels;
* working up the products yielding chemicals;
* generating electricity by burning the solid residue from processing;
* salvaging discharge heat from both processing and from generating electricity.

Projects of this type stimulate the inventors of acronyms. D. W. Horsfall of the Anglo-American Corporation of South Africa suggests the term 'Coalcom' meaning both Coal Complex and also Coal, Oil and Megawatts; in the UK the favoured generic term for multi-product-and-energy systems is 'Coalplex'. Many process variants have been suggested.

But the difficulties already mentioned for power and heat linkage alone, will evidently be greater when further marketing complications are introduced in disposing of material products too. Consequently, it is widely realised that maximum flexibility must be built into such a complex of processes to cope with fluctuations of demand. There is little scope for storing electricity or heat and, though material products can be put into stock, this locks up capital. An appraisal by D. H. Broadbent of the NCB led to the view that the basic requirements for an integrated energy-conversion complex are these:

1. Maximum overall efficiency to conserve energy.
2. Maximum flexibility of output and, as far as possible, product characteristics. Demand for energy varies throughout the day, the year and sometimes from year to year.
3. Waste streams should be made into by-product streams; if SO_2 is stripped from a gas, the sulphur should be recovered, and coal ash should be converted into aggregates.
4. The combine should be capable of acting as an energy reservoir or at least be able to feed into one, such as a gas grid, when demand for its products is slack.

Possible forms of Coalplex

To illustrate both the basic idea and also suggestions for meeting the needs for both efficiency and flexibility, consider a Coalplex for making SNG from coal proposed by Broadbent. The central process is pyrolysis/hydrogenation. The hydrogen for this comes from a fluidized bed gasifier where char from the central reactor is treated with oxygen/steam. There is still a solid residue from this gasification; this solid is burnt in a fluid bed combustor, and used to raise steam. Both the flue gas and the steam are then used in a combined-cycle system for generating electricity. Some of the power will be used in the oxygen plant. If sufficient heat is not produced in the combustor, extra coal can be added to the feed.

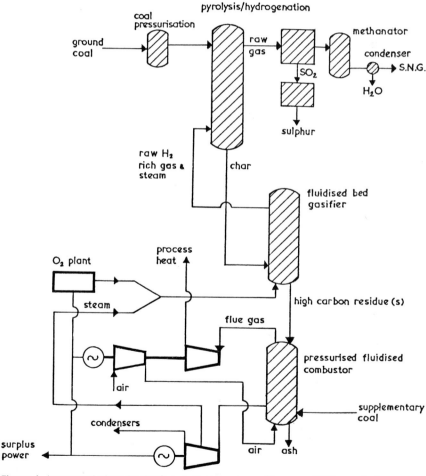

Figure 1. Integrated plant for SNG and electric power. (From an NCB exercise).

When this system is being used for SNG the raw gas from the central reactor will pass to a methanator for fully converting the other gases into methane, as shown in Fig. 1. But if SNG is not required, an alternative reaction route could be made available—not shown in the diagram—to use the gases for making oil products. This would provide flexibility, while the grouping of processes and the interchange of heat and power between units would promote efficiency. Since the design also incorporates fluid bed combustion, there is no need to select coal mining

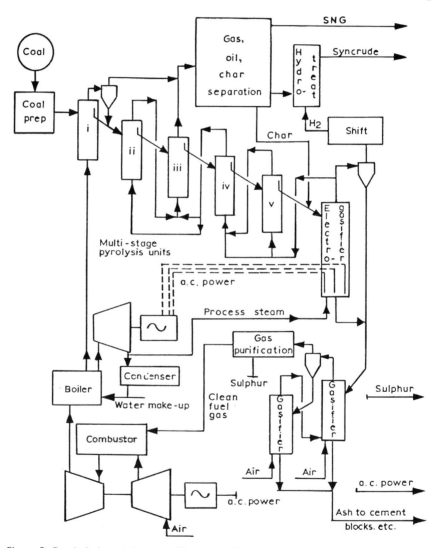

Figure 2. Pyrolysis-based Coalplex. (From an NCB exercise).

conditions to minimise ash in the coal, nor—for the same reason—is the coal likely to need 'washing' to a reduced ash.

Figure 2 shows a further layout proposed by Dr Peter Paul. It is based on the COED multi-stage pyrolysis, briefly mentioned earlier in Chapter 4 (the initials signifying Char Oil Energy Development). In this adaptation coal, after preparation, is fed to the first stage where it is heated to about 320°C by flue gases that have already done useful work. The dried, heated coal moves on to successive reactors where it is heated in stages to higher temperatures; the volatile products thus driven off are collected and separated. Gas may be directly used as a medium Btu fuel or, as shown in the diagram, passes on for methanation to convert it to SNG. Tar is hydrogenated to a synthetic crude oil that can go on for refining. Hydrogen required is generated from char—the residue from the pyrolysis stages—reacted with steam in an electrically-heated gasifier. From this, the further residue of char is burnt in limited supplies of air yielding producer gas ($CO + N_2$) of low calorific value but used directly on site as fuel gas to drive a turbine generating electricity. The hot flue gases from this are used to raise steam for driving a steam turbine also linked with a generator for electricity that can heat the steam/char gasifier; the flue gases are still hot enough to go on to the first stage of heating the coal. And waste streams are duly treated for their potential as sources of by-products.

Two further variants, put forward as suggestions during NCB exercises on the concept of these integrated plants, are shown as Figs 3 and 4.

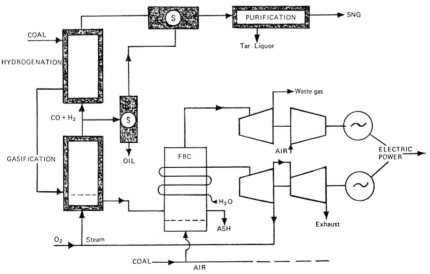

Fig. 3. Coalplex inccrporating gasification and fluidized combustion. S = separator; FBC = fluid bed combustion. (From an NCB exercise).

Designed by H. R. Hoy as 'notional' systems, one has as its main product SNG, while the other is directed to yielding a feedstock for an oil refinery. In both the designer recognises that processing may be simplified and overall thermal efficiency improved if the conversion stage stops short of completely reacting the whole of the carbon of the coal. The residue is passed to a combustion stage to produce steam and electric power. A

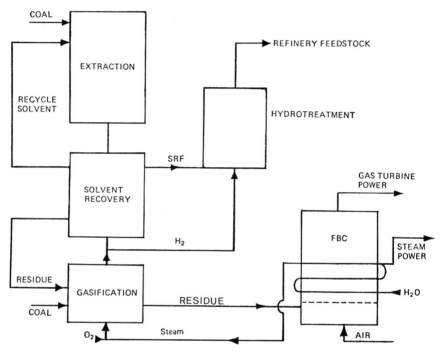

Figure 4. Coalplex incorporating liquefaction and fluidized combustion. SRF = Solvent-extracted refinery feedstock. (From an NCB exercise).

further feature is that this kind of integration is simplified if the residue is burnt under pressure; this avoids the technically difficult problem of designing equipment for reducing the pressure over the hot solids before passing them on for combustion.

Technical and economic studies of these kinds will naturally in due course need experimental verification in pilot-scale plants to provide practical information on the way they operate that could be used in designing commercial-scale plants.

The integrated project COALCOM

A large-scale proposal based on the COED process has been worked out in some detail by D. W. Horsfall for South African coals. Linking it with suggestions for improving the 'washing' of coal (to reduce the ash) his design starts by producing two streams of coal. One with ash below 10 per cent is treated as a source of 'formed coke' after being carbonised; the residue is briquetted and these briquettes again carbonised at high temperature forming a coke that can be used in blast furnaces.

The other stream of coal is designated 'power station coal'. But it is also put through multi-stage carbonisation for saleable gas and tar products that are combined with those from the low-ash coal stream. It is the treatment of the residue that is different in this case. For the high-ash char goes into fluid bed combustion raising steam for generating electricity.

The scale is breath-taking. Horsfall assumes over 15 million tons of usable coal emerging from the washing and going forward in the two streams to their respective COED multiple carbonising stages. The tar made would be almost double that of the first Sasol plant and the plant would generate some 3,000 MW of electric power.

Despite the breadth and sweep of these plans their author points out that no absolutely new technologies are employed. It is the scale and the integration that are new. For formed coke production has been used for many years, the COED pilot plant operated for over 4 years on 25 tons of coal a day and pilot scale pressurised fluid bed combustion has also been successful. The tar for briquetting naturally comes from that formed within the process, consuming some 15 per cent of the production.

How it all fits together is shown in Fig. 5, a flowsheet with figures for expected yields at the various stages. What is missing of course is an estimate of when it will be likely to pay. The original discussion accepts that it would not compete today, but expects this type of arrangement to 'make a good deal of sense' as oil prices continue to escalate and supplies begin to fall short.

COALCON and its difficulties

In the USA, ERDA had intended an integrated hydro-pyrolysis process known as COALCON to be the first of its commercial-scale demonstration plants. It was intended to yield liquid, gas and char as clean boiler fuels and sources of chemical products. A contract was granted originally in 1975 for design of plant to convert 2,600 tons of coal a day, but at the beginning of 1977 ERDA decided that the process needed 're-evaluating' before construction could go ahead.

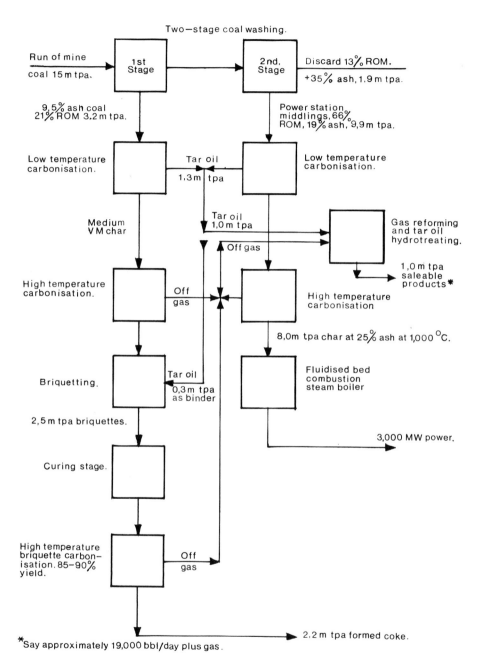

Figure 5. Flowsheet of quantities in COALCOM project proposed by D. W. Horsfall (Coal, gold + base minerals of southern Africa, June 1975 p. 29 et seq.).

In this case the name, looking so deceptively close to that chosen by Horsfall, derives from a coal consortium—COALCON—set up by Union Carbide and Chemico to capitalise on R & D by Union Carbide in coal processing. The main reason given for the delay was 'the high technical risk of the process and its projected marginal economics when scaled up to commercial size'. Reports stated that the test reactors were not more than 6″ in diameter, processing not more than 20 tons of coal a day. Consequently they did not provide enough data for scaling up to the demonstration plant. There was also said to be difficulties with some caking coals—a surprising report since the combined pressurised hydrogen treatment and pyrolysis are in a fluid bed.

During 1977 it appeared likely that the project would continue on a pilot scale but ERDA was pressing both for improvements in design to make the process more economical and for financial commitment by industry.

The point of including this cautionary tale is to indicate the frequent reappraisal of projects that is essential—firstly in considering whether to continue (at any scale) and, more particularly, in proceeding to the expensive stage of scaling-up. In the case of ERDA, this caution continues even after they have awarded a contract for a demonstration plant. The design stage for these commercial-scale plants will inevitably reveal some that should not proceed to immediate construction. As one ERDA official put it—' if none of the design stages for demonstration plants turn out negative, then we are just being too conservative'. The remark was in reply to the group of critics who complain of spending being too fast; naturally there are others who make the opposite complaint.

Linking with nuclear heating

A further stage of integration considered has been to link gasification with nuclear energy as the source of heat. This line of approach seems to be particularly active in West Germany. Dr van Heek and his colleagues at Bergbauforschung point out that in conventional gasification processes only about two thirds of the coal is converted to gas; the remainder of the coal has to be burnt to provide the energy needed. They therefore propose to use process heat from high temperature nuclear reactors (HTR) as a source of energy and transform the whole of the coal to useful gas. The advantages claimed are

* saving coal;
* reduced emission of particulate matter;
* reduced emission of SO_2, NO_x and other pollutants;

* in countries where heat from coal is more expensive than that from HTR, lower production cost for the gas.

Projects under development will use process heat from the HTR in the form of hot helium for steam gasification and for hydrogasification and it is estimated that the process could be available on a commercial scale by 1990.

Prospects

As with nuclear power projects, it seems very probable that the date suggested, by the enthusiasts for the nuclear-linked process, is optimistic. Even for the purely coal-based multi-product-plus-energy refinery it seems likely that because of the complications, completion may be for the UK at least some 15 to 20 years ahead. Earlier plants could well have more limited objectives.

Yet the technical and economic logic of this style of integration is so strong that it must inevitably drive research workers in all the major coal producing countries to study the problems of knitting plants together to gain the great economic prizes that can result from solving them.

Further Reading

'Proposals for National Programme of Development on Coal Conversion Processes'. NCB. Jan 1974.

L. Grainger. 'Future Trends in Utilisation of Coal; Energy Conversion'. Paper to Meeting of Royal Society 'Energy in the 1980's' Nov 1973.

Coal Processing Research. Proceedings of a Symposium on Coalplex held at the Coal Research Establishment May 1973. NCB.

D. W. Horsfall. 'The case for Coalcom'. Coal gold & base minerals of Southern Africa, June 1975, pp 29–53.

S. N. Rothman et al. 'Application of COED Process coal-derived liquids in a petroleum refinery'. Intersociety Energy Conversion Engineering Conference Proceedings, 9th 1974, pp 1027–1034.

Total Energy Conference, 1971. Proceedings. Institute of Fuel.

Conference on Energy recovery in Process Plants, 1975. Institution of Mechanical Engineers.

'Energy R & D in the United Kingdom'. A discussion document prepared for the Advisory Council on Research and Development for Fuel and Power. Department of Energy 1976.

'Philosophy and concepts of coal gasification by nuclear heat'. Dr. K. H. van Heek, Prof Dr H. Jüntgen, Prof Dr W. Peters. 13th World Gas Conference, London 1976.

Chapter 9

The Coal International

There is no international organisation called the Coal International (though there is a UK-based one called British Coal International, discussed below).

Yet there is a network of organisations and a fine tracery of personal contacts on coal research that have the same effect. It is obviously much more economical in time and effort of researchers, and therefore in expenditure, if experimental experience and ideas are exchanged—with due safeguards for possible commercial complications. For though coals do vary greatly from each other in their properties, yet information from experiments on one coal can give guidelines on the behaviour of others found elsewhere.

It is part of the conventions of science and technology that results of a large part of research are published. Both publication of this type in journals and visits of technical men between research institutes in various countries have long been a feature of coal research. But when the European Coal and Steel Community (ECSC) was set up in 1952 it provided a formal basis for international co-operation in research on coal among its member countries. This has been supplemented both by other agreements within other groupings of countries and by special arrangements between pairs of coal-producing interests across national boundaries. These 'interests' may of course be companies or consortia of companies and not only official national representatives.

In Europe the Association of Coal Producers of the European Community (CEPCEO) has set up a technical research committee—which includes an 'observer' from Spain—working separately from the ECSC though well represented on its governing bodies for these researches. Incidentally, ECSC is the oldest of the European Communities; its executive body is now the European Commission.

ECSC

The European Commission supports coal (and also steel) research using funds from the ECSC budget raised by levies on coal and steel production of its members. In 1977 the amount of aid granted for coal research was £10·8 million; of this about 60 per cent went to mining technology and the balance to work on coal utilization. Normally aid is restricted to a value of 60 per cent of the total cost of each project supported. Advising the Commission in choosing projects is a coal research committee with representatives of coal producers, research institutes, miners' unions and universities. Progress is then monitored by expert committees.

In the UK, about half of the work at CRE is supported in this way (although levies from the NCB itself contribute to these funds). Among the investigations specifically receiving such aid were those on gasification, liquid extraction, manufacturing polymers, solution in super-critical gases, production of high-purity carbon and improving industrial coal-burning equipment.

ECE Coal Committee

The Economic Commission for Europe (ECE) is a part of the United Nations Organisation and therefore—unlike the ECSC—brings together the countries of East, as well as West, Europe. It has a Coal Committee dealing with all aspects of the coal industry, commercial and technical. It works by means of circulating reports from member countries, organising symposia, setting up special working parties, organising study tours and annual sessions. Because of its nature, its programme of work tends to be based on national documents submitted by governments—for the UK, by the Department of Energy—to a secretariat at Geneva. Much of the international information I have quoted in earlier chapters has been drawn from these sources. In January 1976 it ran a symposium on gasification and liquefaction of coal. Further meetings for 1977 and 1978 were intended to promote 'intensified co-operation'. The general impression given by the arrangements is that they are attempts to try and ensure that the coal countries outside the more closely-linked IEA grouping (discussed below) do not become isolated from major developments.

IEA Coal Projects

The further set of initials, IEA, is derived from the International Energy Agency—founded in autumn 1974. The Organisation for Economic

Cooperation and Development (OECD) 'took the lead' in setting up the IEA, providing staff ar d accommodation in Paris. The formal relationship is that IEA is an autonomous body within the framework of OECD.

Its objectives, in brief, are:

* to take measures to meet oil supply emergencies;
* to reduce dependence on imported oil by long-term co-operative efforts;
* to promote co-operative relations with oil-producing countries and other oil-consuming countries.

It is worth recalling that OECD has 24 member countries comprising the West European countries with USA, Canada and Japan. Of these, 19 participate in IEA; they are Austria, Belgium, Canada, Denmark, Germany, Greece, Ireland, Italy, Japan, Luxembourg, Netherlands, New Zealand, Norway, Spain, Sweden, Switzerland, Turkey, United Kingdom and the United States.

IEA quickly agreed to develop and implement a strategy for research and development—fostering effective national programmes but also undertaking co-operative projects in high priority areas. Fourteen international working parties, made up of specialists in particular technologies, respectively steer programmes dealing with eight projects based on 'non-conventional' energy systems (such as solar energy, biomass conversion, wind power) and a further six dealing with 'conventional' ones.

Among the latter is the Working Group on Coal Technology. In each of these areas, one country has been asked to take the lead in establishing the group and in servicing it. For coal, it is the UK that has the responsibility and Mr. Leslie Grainger is the chairman of the group. Within it, agreement was quickly reached during 1975 for five projects. They are:

(a) Experimental Fluidized Bed Combustion Plant;
(b) Technical Information Service;
(c) Economic Assessment Service;
(d) World Coal Resources and Reserves Data Bank Service;
(e) Mining Technology Clearing House.

A formal organisation, NCB (IEA Services) Ltd was set up to manage the projects with a Head Office in London, all activities being funded entirely by the IEA members participating in the various projects. In 1977 eleven members were involved in all or some of the projects; the detailed pattern of participation is shown below:

K

Technical Information Service	Economic Assessment Service	Resources and Reserves Service	Mining Technology Clearing House	Fluidized Bed Combustion
Austria	Canada	Belgium	Belgium	Germany
Belgium	Germany	Germany	Canada	UK
Canada	Italy	Italy	Germany	USA
Germany	Netherlands	UK	Italy	
Italy	Spain	USA		
Japan			Spain	
Netherlands	Sweden	Canada	UK	
Spain	UK		USA	
Sweden	USA			
UK				
USA				

At that time, New Zealand was still reviewing plans for its participation and expected to join one or more of the projects in due course.

The Technical Information Service has set up a central storage and retrieval system; it is in the form of a computerised data base with input of abstracts from member countries. The aim is to be able to answer specific queries from members on any aspect of coal technology, and also to prepare critical surveys in areas of particular interest.

The Economic Assessment Service has a role best illustrated by its current programme. Ranging more widely than comparable national studies, the service has projects on the economics of plants for converting coal and for power generation dealing with the way that technical and economic conventions differ from one country to another. Work on cost and availability of coal in the future, costs of preventing pollution and costs of coal-based energy compared with that from alternative sources, is also based on considering different countries.

The name of the *World Coal Resources and Reserves Data Bank Service* clearly defines its function. Covering geology, mining, processing, transport and utilisation, its aim is to provide information in internationally comparable terms. The World Data Bank uses the United States Geological Survey computers with the London office linked into the system for input of data drawn from outside the USA.

The *Mining Technology Clearing House* compiles registers of current R & D, investigates ideas for collaboration on R & D, and brings together personnel working on similar technologies. To overcome difficulties with material that may be confidential, it has a Correspondent in each member country with some discretion on circulating material.

Under the control of the *Experimental Fluidized Bed Combustion Plant* project is the Grimethorpe experimental facility—the largest of the

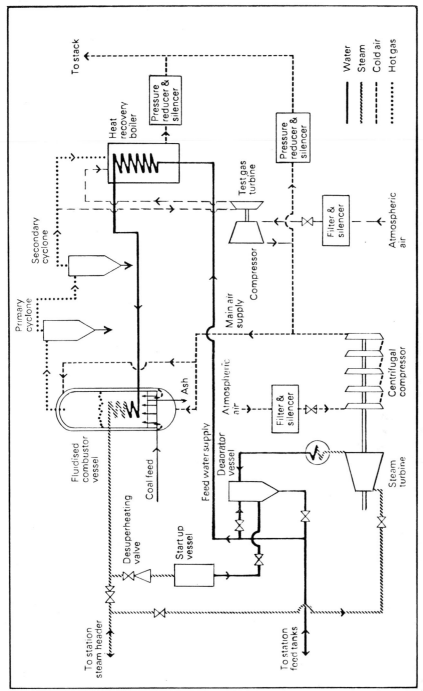

Figure 1. Diagrammatic flow sheet of the Grimethorpe experimental facility. (IEA).

five IEA coal research activities. Due to cost about £17 million over a seven-year period (from November 1975), its objectives are

* to build flexible experimental plant where relevant technical detail for pressurised combustion systems can be investigated;
* on this plant to carry out tests over a wide range of operating conditions;
* from the results to provide design data for large commercial plant.

Fig. 1 shows the diagrammatic flow sheet. The plant is due to be a medium-size industrial-scale unit consuming over 10 tons of coal an hour corresponding to a thermal capacity of 80 MW. Designed to operate at up to 10 atmospheres pressure, it will have a bed of about 2 m by 2 m capable of being used at depths of between 4 and 8 m. The test gas turbine, shown in the diagram connected with dotted lines to the hot gas circuit, is likely to be added in a second stage. In addition to these five projects, plans are in hand for further ones—including pyrolysis research, likely to be placed at Bergbauforschung in West Germany, and treatment of gas liquors.

Other agreements

There is also an International Committee for Coal Research set up to ensure continuous exchange of information among coal producers of 'the Western World' which runs international conferences. Then, supplementing the arrangements made under these organisations are agreements, exchanges of letters or similar formalities to promote collaboration between the NCB and Australia, Czechoslovakia, Hungary, Ruhrkohle AG, Saarbergwerke AG, Poland, the Soviet Union and the USA.

For the Electric Power Research Institute of US, CRE is investigating corrosion performance of alloys for boilers and gas turbines used with fluidized combustion systems (as noted in Chapter 6). COGAS Development Company (also of US) is sponsoring research on gasifying 2 ton per hour of char to produce a synthesis gas with a low nitrogen content as part of the complete COGAS process for making SNG and oil from coal. A pilot-scale rig for fluidised combustion at 6 atmospheres pressure (thermal input 6 MW) is being developed under contract for ERDA. This is intended to provide data for designing combined cycle plant for generating power.

These invisible exports are in addition to the normal export type of activities in which the NCB participates; the latter usually tend to involve joint action in the overseas country where coal-based operations are developing. For these kinds of activities, a new body called British Coal International has been formed. This is a co-ordinating body for NCB overseas units and other joint organisations with private commercial companies engaging in coal mining, coal trading or the areas we have been discussing—combustion and coal conversion.

The programme at the Westfield Development Centre of British Gas, resulting in a successful slagging version of the Lurgi gasifier was funded by a consortium of US companies, with participation of Lurgi of Frankfurt. It is this that has been adopted by ERDA as the basis for one of the demonstration plants for high Btu gas (SNG), as quoted in Chapter 7; the second is the complete COGAS process.

Further examples of this flurry of agreements intended to minimise duplication of effort, are the pair signed between the USA and West German Governments late in 1977. One was intended to harmonise coal liquefaction projects and accelerate technical progress through intensive exchange of information. The other confirmed joint participation in two major projects; they were a refinery for products of coal liquefaction in the USA, and the development of a Saarberg–Otto gasification plant for making a low calorific value gas from coal in Germany. The West Germans also have agreements for jointly-financed work with Australia on petrol-from-coal researches with the objective of producing 2·7 million tons a year of liquid products, particularly motor fuels.

Broadly, those involved now consider that there has been a big improvement in attitude from the former atmosphere of reserve and secrecy. Today it is evident from the web of international agreements that coal interests regard co-operation in R & D as the key to future coal industry prosperity and feel confident that this can be handled without damage to commercial advantage.

Yet, applying the results of the researches on coal both for efficiently using its energy and its chemical potential will depend on the specific conditions of energy supply, demand and cost in each country. What are the prospects for the UK?

Further Reading

'International co-operation in coal research—25 years' personal experience' Dr. Erwin Anderheggen. BCURA Coal Science Lecture. NCB. 1976.

'International Coal Projects.' NCB (IEA SERVICES).

'Energy research, development and demonstration. Programme of the IEA.' OECD, Paris, May 1977.

CRE Annual Reports. NCB.

Chapter 10

Prospects in the UK

At Britain's first Mining Festival, held in 1977 to mark the 30th year of the NCB, one of the exhibits was a 'coal-powered car'—more precisely, an AA patrol car using fuel made from coal. The motor spirit had been made in a prototype, laboratory-scale plant by solvent extraction followed by hydro-cracking and distillation. It had then been blended and track-tested by British Petroleum who reported that 'they had found little difficulty in incorporating the NCB material as a major component in a fuel which the average motorist would find virtually indistinguishable from a regular gasoline'.

And the AA collaborated because of their 'keen interest in the future development of new energy sources suitable for road transport'. So they readily joined in the project 'to show that this liquid fuel obtained from coal does not require any change being made to the conventional petrol engine'.

The aim of the demonstration was to show that 'coal will be able to become a source of motor fuels when oil supplies start to run out' said the NCB Chairman Sir Derek Ezra. No claim was made that the production was economic. But the accompanying announcements referred to the alternative possibilities of using the light and middle oils from the distillation as feedstocks for making chemicals and plastics and of 'vigorously preparing for the day when it will be economic to make liquid fuels and a whole range of feedstocks from coal'.

From this demonstration stage, using a laboratory-scale coal refinery, what is the logical next stage? Scaling up the plant would normally be considered advisable in two stages; the first would be an intermediate-scale demonstration plant, converting perhaps 20 tons a day and costing £10 million. From this it would be possible to design a full-scale refinery comparable in complexity and capital cost to a petroleum one.

In the UK, the government has repeatedly stated its commitment to developing the coal industry, and has backed this with the necessary financial and legislative support. After the coal industry tripartite working group (with representatives of the government, NCB and the trade unions) had included R & D recommendations in their progress report of 1977 it engaged in an examination of proposals for a £60 million programme.

One project was proposed by the Gas Corporation. This was to develop the slagging gasifier process so that it both gasified coal more cheaply and extended the range and quality of coals it can convert.

The NCB have put forward four projects for support. One is the intermediate-scale liquid solvent and hydro-cracking plant mentioned above. They are also suggesting developments in the process using super-critical gas extraction, in rapid hydro-carbonisation and in a combined cycle plant. This would be a pilot plant gasifying 25 tons a day of coal in a fluid bed yielding a low calorific value gas, used immediately to drive a gas turbine, with the exhaust gases raising steam in a boiler to drive a steam turbine.

The policy guiding this working party is to maintain a level of effort 'appropriate to pursuing the technical options suited to UK conditions'. Furthermore, because there appeared to be some element of doubt about the legal authority for the NCB to engage in these activities within the terms of the original Coal Industry Nationalisation Act of 1946, explicit authority was provided in a new Act, The Coal Industry Act (1977). In this, Section 9 gave the Board power to acquire and treat petroleum, manufacture and sell products of petroleum, and treat products of petroleum to render them saleable 'if it appears to the Board that those activities will or may ultimately provide an outlet for coal or products of coal, or may lead to the development of chemical processes or methods, or acquisition of commercial, industrial or technological experience or knowledge which may lead to new or improved uses of coal or products of coal'. And the international collaborative work was covered by Section 11 enabling the NCB to undertake overseas any of the activities it is allowed to perform in Great Britain.

All this is in addition to the progress on the international experimental facility on fluid bed combustion at Grimethorpe. By the end of 1977 all the major contracts had been signed—covering the buildings, fuel preparation plant, instruments, combustor, coal feeding and main accessory plant. According to the schedules, the constructors were due to start erecting in mid-1978 and commissioning would be completed a year later. The four-year experimental programme was planned to cover investigations on combustion and other features discussed earlier.

Government Support

In mid-1978 the government approved a Report of the Working Party recommending immediate financial support from the government for three of the projects outlined above, costing a total of £43 million, and support in due course for other projects costing about £78 million. The projects to receive 'immediate support' were these:

* Liquid solvent extraction with hydro-cracking. This would produce a synthetic crude oil, suitable for refining into petrol, and a solvent that would be re-cycled back into the process. (Cost £16.2 million)
* Super-critical gas solvent extraction. Here the aim would be to obtain an end-product showing promise as an aromatic chemical feedstock. (Cost £14.8 million)
* Composite gasification. This scheme plans to extend the capabilities of the present high-pressure slagging gasifier; the first stage is intended to develop means of handling fine as well as lump coal and the second to couple a further gasifier—based on entraining coal—to the existing developed one. The purpose of the entrained flow/fixed-bed composite gasifier would of course be to produce SNG from coal. But, operating on *air* and steam it could alternatively be made to yield a gas of low calorific value and it is believed that there will be a market for this product. (Cost £12 million)

The further recommendations for support due to lead to pilot and demonstration projects at a later date covered these items:

* NCB and CEGB should be encouraged to carry out a feasibility study of the prospects for low Btu gasification applied to high efficiency electrical power generation.
* Preliminary work in a further IEA project (with Germany and Sweden) on the basics of pyrolysis should be extended to a larger scale as soon as it is possible to make a judgment on its potential.
* NCB with Babcock and Wilcox should jointly prepare a proposal for industrial scale demonstration of pressurised fluid bed combustion for use in generating power. This will presumably be an all-British effort supplementing the contribution to the IEA plant. At 200 MW (thermal) it is due to be a full-scale commercial demonstration plant applying both gas and steam turbines in a combined cycle. It will be capable of operating with wide, variable grades of coal and of absorbing oxides of sulphur in crushed limestone.
* NCB should also be encouraged to continue studies of in-situ methods of extracting the energy from coal underground. This refers to underground gasification but also to further ideas of combustion and liquefaction and using bacteria to break down coal in the seam into methane.

And, finally, in view of the expected depletion of reserves of oil and natural gas, the Working Party decided to remain in being so that they could review from time to time the whole field of coal technology.

Of the total of £43 million needed over a period of eight years for the three immediate priority projects, £32 million were expected to fall due before March 1983 (the period covered by the government's Public Expenditure Survey). The government accepted the recommendation that it should provide two-thirds of the cost over that period; it would then consider further financial support as required.

What is happening in the great wide world?

Energy plans for the UK have to be set in a world context—as the first Working Document prepared for the new Energy Commission acknowledges. For this world view, the most authoritative source is clearly the World Energy Conference. What is their assessment for the period up to the year 2020?

Oil: For this fuel, now representing about 45 per cent of the world's energy consumption, they broadly confirm 'conventional wisdom'. By the year 2000 they expect new discoveries only to equal present day consumption. No further major discoveries are expected on the scale of those in the Middle East. Because of the wide range of problems in developing oil from unconventional natural resources, the necessary 'enormous investment' will call for government assistance.

Gas: Though the gas industry is optimistic, the international experts indicate that they are sceptical about the possibility of gas playing a more significant part in satisfying world demand outside the centrally planned economies.

Coal: There is no shortage of coal. The doubts are associated with the rate of growth of output. 'There is a noticeable lack of will' they write 'to provide the long-term investment and determination which the development of the resource requires.'

Uranium: Though there could be enough uranium available, the recommendation here was to look to the fast breeder to reduce world requirements. Evidently responding to world anxieties, they added a recommendation that governments should study how to use plutonium 'with the least damage to mankind and to the environment'; the alternative was a substantial reduction in the standard of living.

Wood: As populations increase, forests are disappearing with disastrous rapidity. Increase in oil prices has aggravated this tendency in developing areas.

Renewable Energies: Solar, geothermal, fusion and other renewable sources are seen as doubtful or small contributors within the period.

Generally, they stress the importance of work on conservation of energy and of taking early long-range decisions to avoid later energy shortages.

And, as a general philosophy, the energy experts found it of great value to work together globally and strongly urged 'true international collaboration' in order to give to succeeding generations an overall increase in the standard of living.

The Association of the Coal Producers of the European Community (CEPCEO) in their forward review for EEC countries made only a brief reference to coal's potential as a feedstock for gasification and liquefaction plants. After detailed consideration of probable developments for each fuel, their emphasis was rather on ensuring that nuclear power and our indigenous coal met future needs rather than coal imports which would simply replace one form of energy dependence on imports (oil) by another (coal).

UK Future

All observers agree that the UK is exceptionally well-placed—among industrial countries—for future energy supplies. We have oil and gas, probably for a generation, and supplies of coal to last us for centuries. Of the major established sources of energy, we lack only uranium.

The importance of relating UK policy to the world review is that the latter shows how little we shall be able to rely on outside supplies of oil and gas when those from the North Sea will be in decline. All reasonable assessments stress that imports of oil will then be more and more costly as total supplies also decline; for gas there do seem to be differences in interpretation, but they only delay the inevitable by some years. The Working Document for the Energy Commission concludes 'On most views of the future, the needs of coal, energy conservation and nuclear seem inescapable. Renewable sources could also make an increasing contribution.'

Its forecasts for energy demand are based on two assumptions of economic growth:

UK Energy Demand—million tons coal equivalent (mtce).

	Higher growth			Lower growth		
	1975	1985	2000	1975	1985	2000
Energy	315	375	490	315	350	390
Non-energy	25	40	70	25	40	60
Total primary fuel	340	415	560	340	390	450

The corresponding indigenous energy supplies in 2000 'might comprise':

UK Energy Supply.

	mtce
Coal	170
Nuclear	95
Natural gas	50–90
Indigenous oil	150
Renewable sources	10
Total	475–515

This points to a further highly relevant issue for any discussion of the prospects of a successful oil-from-coal industry—will the supplies of coal be available? On this, the Working Document comments

'Depending on many factors, including the course of world supply and demand for oil, we may find that we need to do more than now appears practicable, either to increase production of coal, nuclear or renewables, or to promote the conservation of energy. If we are unable to do so, the alternative could be an enforced reduction in economic growth, and hence in living standards.'

So the issues we have been discussing earlier—of technical progress resulting in a competitive process for converting coal—also interlink with those of producing enough coal competitively. On both aspects the industry appears to be reasonably confident for the future.

Fuel shortages will push the USA into full-scale conversion plants much earlier—probably operating within the 1980s. But for Britain, the 'best informed estimates' appear to be that we shall have commercial coal refineries early in the 1990s. For that is the time when it is probable that both oil and gas prices will have risen to appropriate levels, and the full technological development of the selected best processes should have been completed.

From that time on, coal sources should increasingly take over the many roles that oil sources now fill in supplying fuel for industry and the home, chemicals for all our many purposes and even the precious liquids that drive our lorries and cars.

For all the indications are that as we move towards the turn of the century, King Coal will again be returning to power.

Further Reading

'Coal for the Future. Progress with "Plan for Coal" and prospects to the year 2000'. Department of Energy, 1977

'Coal in the UK'. Fact Sheet No. 4. Department of Energy 1977.

'Working Document on Energy Policy for the Energy Commission'. Energy Commission Paper No. 1. Department of Energy, 1977.

'European Coal 2000'. A report by CEPCEO, Association of the Coal Producers of the European Community. June 1977.

'Energy Policy Review'. Energy Paper. HMSO, 1977.

Statement by Mr Eric Ruttley, Secretary-General, World Energy Conference, to the Colloquy on Energy and the Environment held by the Council of Europe, November 1977.

Reports of the Conservation Commission of the World Energy Conference, 1978.

'Coal Technology. Future developments in Conversion, Utilisation and Unconventional Mining in the United Kingdom'. Report of the Working Party on R & D. Department of Energy. 1978

Index

A

ACORD   1, 11
Approval scheme for domestic coal
    appliances   68

B

Basic science   109
British Coal International   125, 130
British Gas   102, 131

C

Chemicals, consumer products   14, 25
Chemicals from coal   18, 46, 93, 107
Chemicals, growth rates   14
Chemicals, inorganics   16
Chemicals, organics   18
CO_2 acceptor process   103
Coal carbonising   47
Coal classification   39
Coal composition   35, 93
Coal, extraction with gases   95, 99, 134
Coal, extraction with liquids   95, 99, 134
Coal, gasification   51, 102, 134
Coal, hydrogenation   95, 101
Coal Industry Examination
    Reports   12, 133, 134
Coal origins   33
Coal processing   48
Coal pyrolysis   56, 91, 95, 98, 134
Coal rank, chemical significance   44
COALCOM   121
COALCON   57, 121
Coalplexes   115
COED   56, 119
COGAS   130
Combined heat and power (CHP)   70
Cracking   21
CRE   55, 72, 91, 99, 101

D

Distillation   20, 48

E

ECE Coal Committee   126
ECSC   126
Economics of conversion   110
Energy Commission   135
Energy consumption, UK   7, 136
Energy consumption, world   2
Energy, efficiency of use   74
ERDA   56, 90, 101, 102

F

Fluidization   78
Fluidization, for burning coal   81
Fluidization, at higher pressure   88
Fluidization in combined cycles   89, 134
Fluidization, non-coal applications   79, 91
Fuel cells   72
Fuels from coal   93

G

Grimethorpe Experimental Facility   90, 128

H

Heavy natural gases   23
HYGAS   101

I

IEA   90, 126
Import costs for oil   10
International agreements   130

M

| Magnetohydrodynamics (MHD) | 74 |
| Metals from coal | 53 |

N

| Nuclear heating | 123 |

P

| Petroleum chemicals (petrochemicals) | 18 |
| Platforming | 22 |

R

| Reforming | 22 |

S

Slagging gasifier	102
Smog	59
Smoke eater	67
Solid fuel room heaters	66

U

| Underground gasification | 104, 134 |

W

| Waste heat, using | 69 |
| World Energy Conference | 135 |